I0020016

Artificial Intelligence

Learning Automation Skills With Python

(How to Gain Insight and Build Innovative Solutions)

Darlene Baldwin

Published By **Elena Holly**

Darlene Baldwin

Artificial Intelligence: Learning Automation Skills With Python (How to Gain Insight and Build Innovative Solutions)

ISBN   978-1-77485-816-5

No part of this guidebook shall be reproduced in any form without permission in writing from the publisher except in the case of brief quotations embodied in critical articles or reviews.

Legal & Disclaimer

The information contained in this ebook is not designed to replace or take the place of any form of medicine or professional medical advice. The information in this ebook has been provided for educational & entertainment purposes only.

The information contained in this book has been compiled from sources deemed reliable, and it is accurate to the best of the Author's knowledge; however, the Author cannot guarantee its accuracy and validity and cannot be held liable for any errors or omissions. Changes are periodically made to this book. You must consult your doctor or get professional medical advice before using any of the suggested remedies, techniques, or information in this book.

Table Of Contents

Introduction

People think the notion that Artificial intelligence is connected to those who think that artificial intelligence is linked to Sci Fi dystopia, but it is changing of perception, because artificial intelligence is advancing rapidly and is becoming more integrated into our daily lives. Nowadays, almost everyone is aware of Artificial intelligence. The idea of Artificial Intelligence was first discussed but it took several years to be accepted and to incorporate it into the world of practice. The significance of Artificial intelligence is not to be ignored in the business world. It is a fact that everyone uses Artificial Intelligence in some manner. Artificial intelligence has a prominent place across every industry. With the growing awareness of AI and its application in industries and businesses is now essential for success and even for gaining an edge in competition. The book discusses the way that AI is advancing machine learning and are being utilized to enhance the business. Before we look at the various applications and benefits of Artificial intelligence and machine learning in the first place, we will look at the fundamentals of Artificial Intelligence and Machine learning.

Chapter 1: Getting Started

In the past There was no awareness about Artificial Intelligence among the people and the mention of the phrase in everyday lives was seen as an irony.

In the modern world of business in the present, Artificial Intelligence holds a significant position. Artificial Intelligence plays a vital function and is now essential to the digital transformation businesses have employed to get a an edge over their competitors. AI allows companies to access the massive and ever-growing database.

The question is what caused this change? The answer lies in the revolution that is Big data. The enormous database led to greater research into the methods used to process data and analysis as well as its applications. It is widely acknowledged that machines are more efficient in this respect they are superior to human beings, AI requires machines to be taught to perform these tasks in a way that is intelligent.

The subject gained more attention in the academic-industrial and similar communities. It required more research and the collection of data. This led to a variety of innovations that are creating significant changes around the globe.

What's Artificial Intelligence?

The definition of Artificial intelligence has evolved over years, however the basic idea remains that Artificial Intelligence refers to the machines that can operate like human.

Humans are able to collecting information through observing their surroundings. They then use it to act and cause an alteration. This means that humans could serve as models in the design of AI machines.

It is Artificial intelligence is divided into two major branches based on the development and research work.

Applied Artificial Intelligence:

It is based upon the implementation of the fundamentals that enhance the human brain's ability to accomplish a specific goal.

Generalized Artificial Intelligence:

It is the process of incorporating of machine intelligence, which lets machines complete various tasks by changing their capabilities in accordance with the demands for the job in an ethical way exactly the same way that humans complete the task.

Numerous studies in the area of Applied Artificial Intelligence show that quantum physics is used in AI to predict the possible behaviour of systems. Artificial Intelligence is employed in a variety of industries within the finance sector to evaluate the risks of external fraud , and protect the business from fraudulent activities. It assists companies in predicting the requirements of customers, which leads to more efficient service. Moving towards the manufacturing industry, Artificial Intelligence is used to monitor the process and control them, as well as aids in identifying any flaws during a process prior to.

Moving towards the consumer market We are employing AI technology on a every day routine. Artificial Intelligence is used in the smart phones that use google's Google Assistant and Apple's Siri

Assistant. In the automobile industry, cars that have Artificial Intelligence Artificial Intelligence will take a control over manual vehicles.

While AI isn't as intelligent and efficient as the human brains, it is predicted to behave as a human brain, with increased computing power and brain-like abilities as it develops and launches of the technology for computers known as a neuromorphic processor. It is still in development and will eventually execute the brain simulator code with efficiency.

The most important developments in AI

The fundamental principle behind AI's operation AI is to mimic the humans' brains.

The field of research that has been a success in this time is" machine-learning". The book will discuss this in detail. "Artificial intelligence" along with "machine training" are so similar to one another that they are used interchangeably terms.

But, this isn't the case. Machine learning is a term used to define and discuss the current state of the art within the realm of Artificial intelligence.

Machine learning is distinct from AI because it is related to the concept to program the system in a way that it doesn't depend on humans for instructions in every step but utilizes its own abilities to determine the next move.

Machine Learning Defined

The most common conclusion drawn from Google's self-driving cars and Netflix's recommendations is that the system makes use of machine learning in some degree to enable the system to autonomously complete the task it done before in the same way it has previously performed without needing to be re-directed by humans to perform the execution of the identical task.

Machine learning is the process of designing machines in a manner that it acts and thinks as humans, but with no human interaction. Machine learning systems collect data about learning through interactions with the world and from the data transmitted to the machines.

Although machine learning only came into existence in the last few years, the fundamental algorithms for pattern identification were in use for a long time.

But, the machine learning models are more closely linked to the complicated datasets and learn from previous calculations and forecasts to help make important decisions recently. A company can stay clear of risky situations and earn profit by developing the right model using AI or machine learning.

Machine Learning Tools in Business Sector

A few of the most famous firms in the world are employing machine learning techniques to enhance their products. This includes Google, Microsoft and Apple. To speed up the performance of Artificial intelligence on the iPhone, Apple has produced its Core ML API. Azure cloud services provided by Microsoft comprise an Emotion API that detects emotions of humans like love, anger displeasure and even surprise.

They are created to be able to are able to adapt to changes in the rules, most recent information, and then operate in line with the modifications.

The most basic tool companies utilize to handle the vast amounts of data and make big decisions is machine learning. The machine learning program allows companies to be more vigilant and observant, and to stay ahead of threats that are present human error, completion and.

The business must think about the security of their data as well as employees' morale staff when applying machine learning. A huge amount of data is analysed using Machine Learning. It also provides companies with more rapid results in order to choose which options are the most lucrative among the possibilities to make profit, but lots of time and resources are used in the process of training large data sets. Combining machine learning with Artificial Intelligence and Cognitive technology will make it more efficient in the analysis of large amounts of data.

Chapter 2: What Is The Reason For Use Of

Autificial Intelligence?

It is believed that, among all the advancements, Artificial Intelligence is going to make our future more sophisticated and safer. The speed of progress of AI is astounding. The wonders of AI are evident everywhere as evidenced by the vast amount and diversity of information available, cost-effective data processing and storage. While machine learning and Artificial intelligence have been around since the middle to the end of 20th century it's been more widely accepted in the last few decades. It is a question of why machine learning is so crucial today?

Machine learning is an field of computer science. It can help computers behave and behave like humans. The goal is to design algorithms that enable machines to learn to perform the requirements of a job.

It may be linked to the field of statistics as well as mathematical optimization. It is comprised of many techniques like Supervised learning, Semi-supervised learning, Unsupervised learning, and reinforcement learning, each of which has their own algorithm and applications.

What is Artificial Intelligence?

If you do not take care of Artificial intelligence, you'll be left behind by your rivals. It isn't too long where technology will transform the world like magic.

The speed at which AI is gaining acceptance across the world is evident. It will surprise you to find out that in 2015 Google created an agent that does not just communicate with employees, but also conducts itself ethically when responding to inquiries and also share its thoughts.

In 2015, a new agent was created in The Deep Mind which surpasses the humans. It was able to receive game pixels and points as inputs during the rate of 49 Atari games. It was the Depp Mind broke its own record by coming up with an entirely new gaming device called A3C in the following year.

Human beings also were defeated by Alpha Go. Alpha Go in the chess game that humans were regarded as champions of. It was a bit difficult for human champions to know it was difficult to understand how machines could beat the complex Chinese war strategy.

Let's explore some instances in which machine learning is used to better understand the applications of machine learning. Machine learning can be seen in many applications such as cyber fraud detection, a self-driving Google car, and online suggestion engines that are available on social media sites such as Facebook and Instagram lists of the latest films that are shown on Netflix as well as on Amazon, "Identify something for yourself" and many other pop-up suggestions of items based on your preferences.

The examples above illustrate that machine-learning is a necessity in the current world, which is becoming very data-rich. The machines are constructed to sort the needed information, which is utilized in a variety of advancements.

We are witnessing a continuous increase in the use and benefits of machine-learning and the need for it. In the present, there are vast amounts of data which are complex , and machine learning is utilized to deal with the data. Alongside the ability to comprehend machine learning extracting data as well as its results are also made simpler and speedier to be competitive.

Chapter 3: Supervised Learning

How much profit can be anticipated in the event that a business spends more on advertising via digital media of its services or products? What are the predictions for the stock market in the next day?

The difficulties in supervised learning process are triggered by an array of data that contains examples of trainings together with marked labels. As an example, when someone learns of how to classify numbers written in handwriting an algorithm for supervised learning will take thousands of images of handwritten numbers, as well as names that contain the right number every image represents. The algorithms reveal the connections between the pictures and the figures that are related to them to them, and then use this learned relationship to categorize the most recent pictures that are also fresh for machines.

The function of supervised learning is understood by looking at this scenario. Let's predict the annual income of an individual on the basis of his education. A model must be constructed to calculate the connection f between number of years of educational experience as X and the level of income Y.

To make a more complex model, we could add new rules that show the kind of educational level, the rank of the school , and their experiences over time.

It could also be that the rule-based programming could fail to function in the appropriate manner with the information that is complicated.

This issue is tackled by the machine learning supervised that lets the computer take over the task. It makes decisions that may not be flawless, however it will at least solve the problem. The main difference between machine learning process and human learning is that the latter occurs on a computer's hardware and that statistics and computer science are involved. Machine learning is a process of learning to connect the level of income and education degree of a person starting from zero, making use of learning algorithms for generating the task of labeling training data. The function that is learned is utilized to calculate the income of the person who has an income of Y. Years of education X is a known and used as an input. This means that the method is then applied to test data that is not labeled to calculate the value of Y. The purpose of the machine learning model is to arrive at the exact value of Y, when various amounts of X is provided.

Two purposes of supervised learning Two functions of Supervised Learning: Classification and Regression

Regression functions are employed to calculate the value of the continuous target variable in relation to various known values of an input variable X. For instance, it could determine the expected lifespan of an individual using an X factor.

In this situation the variable that is targeted assists in the calculation for the variable that is unknown. The unknown variable is only determined by using values of X. the continuous proves that there aren't any gaps in the values of X. Weight and height of a person are considered to be constant values while discrete variables are able to be used for a certain amount of possible values. An example of a discrete variable would be the number of children that a person has.

The projection of the income of a person is an excellent illustration of regression. All relevant variables such as the amount of years of education or designation of work or work experience constitute the input variables X. These characteristics are either categorical or numerical and are often referred to as "features.

To calculate the relation that f is between X as well as Y we need full training data on those features as well as their relationship to the Y. Data is composed of test data as well as training data. Data labeled with a training symbol is data for training and unlabeled data is test data. This is difficult to predict the value of the value of Y. The model has to be applicable to completely new circumstances for it to work using test data too.

What's the method of making accurate models to create the needed predictions for the real world? The answer lies in the application of the algorithm for supervised learning. The next topic we're planning to go over is" getting to know the basics of algorithms". The methods to apply categorization and regression will be taught and machines learning concepts will be explained.

Linear Regressions in Machine Learning

The following is a simple regression model where the actual value is used as the variable to be used as the target.

Let us begin with an illustration. Imagine that we have a collection of data that contains information regarding the location and cost for the home. We are trying to build an algorithm that can determine the cost of the property. The following is the list of information:

Area (sq. feet) Cost (1k$s)

3456 600

2089 395

1416 232

When we set our data on graphs and it appears as follows:

Let's take a closer look at the linear regression in order to better understand:

A linear regression can be thought to be linked with the model that is linear. The relationship between input variables as well as the single output variable, Y are identified by the model. A proper combination of input variables or X values is essential for this calculation.

One linear regression happens the case when there is only one input variable. And when multiple variables are involved it is known as a multi-linear regression.

Simple Linear Regression

Simple linear regression illustrates the connection between the goal variables and input variables as a result of a regression line. Utilizing the equation y=m*X+b the range of variables is represented. This equation says:

X stands for independent variable

Y stands for dependent variable

Intercept is demonstrated by the following video.

Gradients are shown using the symbol m

Multiple Linear Regression

If there is just one input variable, it is possible to use the one-liner regression formula. may be utilized. Most often, you'll have to handle the large set of data that has numerous input variables. Multiple linear regression can be utilized to data with more one elements. By modifying the formula for linear regression with a simple formula we can obtain the formulae for multiple linear regression.

In the multiple linear regression, the hyperplane of an area with n number of dimensions is considered for predictions. If, for the dimensions is 3 the plot will appear in the following manner:

Cost Functions

Different lines are generated by using different weight values . Our goal is to determine those weights that work most effectively. The question is what is the best way to find the line of data that provides the greatest fit, and, if you are given a two lines to choose from, what will you do to determine which is the one that best fits? First, you'll develop an cost function which can determine the distance between Ys and the value of Xs if an estimate of the value of W is provided.

20

What is the best way to determine the target value be calculated for a certain set of weights?

We will employ the cost of error function mean squared within linear regression. It is the measure of squared errors that lies between the forecasted value and real value for the data locations (xi, Yi)

Residuals

The cost that is calculated through the function of cost and is how far between actual desired target and the forecasted one is referred to as the residual. The term "residual" can be defined in the following manner:

These residuals can be greater in the case of a line that is not in the direction of the ends and also the cost functions too will be more expensive. However, the residual is reduced if the range is too close to the ends and the same is true for Cost function.

Utilization Cases for Supervised Learning for Data Science for Business

The methods and procedures used in business have been completely changed through the use of machine learning. The primary thing that's distinct in ML and other technologies that automate is rule-based programming. The engineers are motivated by ML algorithms that control the data but without the need to programme computers to use various routes to solving problems. The machines are however capable of making decisions based on the data fed to them. The businessmen were required to review their decision-making methods due to this ability of the machine

In simple terms of a typical man who is ignorant of the machines, machine learning process is utilized to evaluate the outcomes of the feed data. For example, in the case of an online store the life-time value of a client is determined by the successful relationship that the customer has with you in the future. It is possible to use the machine learningtechnique when you have the record of a customer's visit as well as interactions with your website. This will also give you the most recent details about customers that will yield both in terms of financial gain and an ongoing relationship with the company.

Supervised learning is one of the most popular learning method. In this section we will discuss the supervised learning method and its ease of access for any business that plans to focus upon learning through the ML program. We will also look at the relevant cases to this.

How Do Controlled Machine Learning Operate?

It is essential in the an environment of supervised machine learning that the solution to the problem is not unique to the previous data, even if it's new to the data that has been fed. It indicates that there are higher likelihood of finding correct answers from the historical data , and we employ algorithms to identify these in the newly compiled dataset. For instance, in a set of public information gathered by an Portuguese bank during the year 2012 , in its marketing program, the institution focused at helping its customers sign their consent to deposit terms through phone calls and promoting the service.

A data set is comprised of columns and rows that are which are laid out in the form of tables. These data elements are stored in the rows and the columns hold the variables. The values that are likely to be anticipatory for data are referred to by the name of target variables. They are are incorporated in the data sets with labels. The target variable of the data set will inform you whether the client was able to accept the deposit terms after receiving the call, or not.

Supervised Learning Cases: Use Cases

Tech Emergence in 2016 published the results of a brief survey conducted by the experts in Artificial intelligence. Participants in the survey discussed readily accessible machines learning applications for large and small-sized companies. Participants were able to vote repeatedly, however they didn't feel that it was required since the results were already evident.

It is vital to remember it is important to note that Tech Emergence failed to clearly group the use scenarios among the various ML tasks. For instance, Big Data showed relevance to more than one group in the study because almost all kinds of data is handled by the algorithms regardless of the field of study it is related to. In addition, marketing and sales tasks at times can be misunderstood and be discussed and used within the same field

By using the ML that we have used for this set of data we can calculate the probabilities of subscribers from other customers of the bank.

The process of training an algorithm ML involves the selection of data to be input into an algorithm with the aid of a mathematical process. This results in an algorithm that defines the variables to be used in the subsequent data. In this case the algorithm must be able to split data into two kinds, whether it's a yes, or no. The supervised learning process is centered around three main areas:

Binary Classification

An excellent example of binary classification can be seen in the above illustration that shows how an algorithm divides the data set into two major divisions.

Multiple classifications

If there are multiple classifications the algorithm must be used to choose the appropriate response for the target variable from a pool of responses.

Regression

The value of a constant is predicted by regression models whereas models of classification aim to forecast categorical value. For example, forecasting the average profits of a customer in relation to the longevity of a client is considered to be a straightforward regression problem. Problems to be emphasized for the purpose of supervised learning

Data Collection

Machine learning is built on data. It is possible to make the most error-free models when you have a huge amount of data. We will help you learn more about data collection.

Labeling Data

It is possible to conclude from the last instance that labeling isn't in any way difficult. It simply requires the proper data collection, followed by the label classification at the end of the call or campaign. It is, however, easier to do than say.

Let's imagine a scenario where a person needs to separate freshly picked apples from the damaged. First, you will need to prepare an assortment of pictures of rotten and fresh apples If you decide to utilize a machines to separate them. The next step is to label the apples. Labeling and image recognition can be lengthy as the machine must scan through thousands of images to differentiate between fresh and the rotten apples.

in 2006 Google began to make the image crowdsourced labeling game by developing games that involve the labeling of goods and inviting users to join in on the fun; which led to the further advancement of Artificial Intelligence of Google. Similar to this, Amazon has also created a platform known as the Turk platform, which offers a possibility to users to earn money from allocation of data labels.

Marketing and Sales and Marketing

When we discuss Machine learning in the field of marketing and sales digital marketing, the digital marketing and growth in sales through online channels will be discussed. Internet users use it to communicate with each other, which allows sellers to track the activities of their customers. While unsupervised learning has numerous applications in the fields of marketing and sales, it's not as effective as the huge effect of the field of supervised learning.

Human Resource Allocation

To learn more about different aspects of HR machine learning is employed to retrieve data from HR software and also information related to holidays of employees and vacations. This data provides a number of forecasts. A majority of auto firms are able to forecast about the absences of their workers on certain dates and also their extended time off.

Time-Series Market Prediction

The time-series market prediction the principal field of machine learning where the statistics will describe the events that are based on the timing. Examples include the shifts in market statistics that occur in time. Because it's concerned with time, it also forecasts seasonal shifts and holidays.

In today's business, time-series information is utilized within the company to develop better services to the user interface in order to keep and attract customers. For example, e-Commerce website can avail the information regarding Black Friday to predict consumer behavior during that particular period and provide discounts and sales to attract buyers in line with.

Security

Most often, unsupervised learning methods are employed in cyber-security processes as they can spot irregularities and data sets that could cause harm. There are also instances where the application of the supervised learning method proved beneficial.

Spam Filtering

According to reports, in 2017, the majority of the emails were deemed in the spam folders. This high volume of spam-related emails underscores the importance of metadata and textual filtration to aid in detection of spam.

Malicious emails and Connections

To ensure the survival of nearly all IT offices , the identification of malware is essential. Nowadays, public sets of data provide details of malware, which are identified and used to develop models for simple identification of spam.

Asset Management and IoT

Digitalization has pushed the boundaries of even the infrastructure within IT. Companies are getting more sophisticated through the application of the internet-of-things. They make use of various smart sensors to gather data and transmit the information to cloud servers accessed by public , resulting in centralization of data and its use in the management of resources and expansion in supply chain.

Logistics

The logistic managers are accountable for the timings of delivery. Machine learning is employed in logistics to discover the most suitable solutions. It also displays the records of budget allocation along with the characteristics of the driver and other information that are evolving. One of the biggest issues facing organizations today is managing supply chains, especially for those that hold real-time information sets and resources. In this context it is crucial for these organizations to develop an AI-based recommendation systems that provide the flexibility to modify.

Prediction of outage

Machine learning is also used to detect the flaws that can hinder the smooth functioning of a machine through studying the records of past events of outages on machines. Modern ML algorithms can predict things that cannot be anticipated by humans. This allows a company to reduce the expense for repairs to machines or the stoppage of production due to damage to the machine.

Entertainment

Machine learning is also used in the realm of entertainment. Users are connected to machine learning algorithms in real time. Visual alterations and face recognition are also a part of. A majority of companies working in the entertainment industry have only recently begun using the AI for the exchange of software, which can be utilized in different products.

Jumpstarting Machine Learning involves Data

Supervised learning is utilized in fields where they generate a large amount of data. The data is centralized and organized within the organization internally. When the information is labeled it makes the work simple.

The data is getting more machine-readable as companies are being transformed into digital. This has led to the removal of paper ledgers and spread sheets from official process and the rise in the usage of various tracking softwares and CRM to keep track of the information. The initial step in machine learning process is to examine the data before deciding on the most appropriate classification plan as well as the terms used in regression to obtain the corresponding responses.

Chapter 4: Unsupervised Learning

The data isn't labeled in the case of unsupervised learning. It is possible to reconstruct the input data using the help of a representation even when it isn't labeled. There are many opportunities for the unsupervised learning process on unlabeled data based on the small proportion of data that is labeled all over the world and the idea that supervised learning isn't appropriate for the majority of data, and the idea that the data that is trained aids in the optimal learning of models. It is claimed the next generation of AI will be based on unsupervised learning.

It is possible to incorporate neural networks in classification, clustering, and regression. They are universally used approximates to make use of the non-linearity. After pre-training, they produce the "superior" features and also acquire the technique to compile data. Neural nets use random loss functions to facilitate the integration of inputs and outputs.

Different kinds of algorithms including the traditional ML algorithm may benefit from the properties generated by neural networks. These algorithms may incorporate data, logistics regression, used to classify input and simple regression to make predictions.

They are thought of as the components that link to other properties. In the case of the help for Image Nets, for example Image Net, a person can build a neural system to comprehend the characteristics of an images. This is achieved by the technique of supervised learning. Through the use of the neural network, you could also acquire the features and alter the algorithms in a second one to classify images.

Here are some of the aspects of the usage instances that are created by using neural networks.

K-means Clustering

This algorithm is based upon the idea of unsupervised learning. It automatically labeling stimuli based on the raw distance between stimulant and the inputs within the vector space. This algorithm does not come by target or loss functions, but it does have centroids. The centroids are generated by regular averaging of data points. The new data is categorized according to its distance from a particular centroid. Each centroid is identified.

The creation of groups of data points is the primary function of clustering.

Steps of Clustering K-Means

1. The definition of the k centroids that start randomly. are also an alternative to methods to start the centroids that provide greater ease.

2. Find the nearest centroids, and take note of the clusters. Allocate each data point to the clusters of k. Nearly all numbers are allocated to the closest centroids group. Hyperparameter is the measurement of the proximity.

3. The centroids are moved to the midpoint of the clusters. The most recent location of the centroid is determined by the average for all the data points within the group.

It is suggested to repeat the steps 2 and 3 for as long as the central point continues to move.

The process of clustering k-means is explained in a brief manner.

The handwritten numerals are defined by clustering of k-means that occurs on the ground.

Hierarchical Clustering

Hierarchical clustering is similar with regular clustering. There is only one difference: in the order of these clusters. It's used to reach the desired amount of clusters. Consider, for instance, the layout of the products on the site on Amazon, which acts as an online marketplace. The products are arranged in diverse categories so that you can efficiently search the items, however when you search for a particular product, you may need more detailed information. If you use algorithms, cluster assignments create a fantastic chain of command that informs the business about the various types of clusters. Additionally, you can define what number of clusters that are required.

Methods to Hierarchical Clustering

* Begin with N clusters, every data point is part of one cluster.

* To create the N-1 groups, mix the two clusters close to one another.

* You have to determine the distance between the clusters once more. There are a variety of ways to accomplish this. One method is called typical linkage clustering. This technique will need to consider the distance between 2 clusters as the distance between all numbers , and vice versa.

* If you do not discover one cluster of N numbers of data elements, you will be instructed to repeat the first three steps. The dendrogram will appear as an erect tree.

In the dendrogram, you will need to choose various clusters, and then draw an horizontal line. In this case, if there is a need to have two groups with k=2, you'll have to draw the horizontal line from the distance.

Curios Pupil with Unsupervised Learn

Since the last few years, machine learning technique has gained significant acceptance across various fields. The key reason for this popularity is the creation of neural networks. In both of these types of approaches the training signal generated by a human being is transferred an computer. This means that the learning is managed by a human.

According to some researchers, extensive education is the only thing needed to develop general intelligence. Others think that the practical application of knowledge demands more individual approaches to learning.

For instance, a toddler may learn if presented with something which is presented to the child as a reward to complete his task.

This is the same for unsupervised learning in which the agents are compensated by learning something new about the information they have access to with no specific goal. This helps in the development of autonomous intelligence . It can be described as learning solely for the goal of learning.

The most effective reason for unsupervised learning is that while the data that is transferred to the algorithms that learn have a rich structural structure internally, the objectives and rewards that are presented during training don't require much complexity. This suggests that an over amount of knowledge gained from algorithms requires knowledge of the data by itself, instead of making use of the data from the general job.

Learning through creating

The principal purpose behind Unsupervised Learning is that it helps help the algorithm in coming with its own unique kind of data. It is referred to as a an generative model. This model does doesn't replicate data, but instead presents a model of the class that is used to determine where the data is obtained. An authentic data instance is derived from models that are generative. Insofar as images are involved, the most suitable model to be constructed can be described as one called the Generative Adversarial Network (GAN) where two networks, one being a generator and a discriminator, are involved with the procedure of discerning.

Improvements are made to pictures that result by repeated processing . The constant activity of the network produces the most realistic pictures that appear as if they were taken from actual photographs. Furthermore, Generative adversarial networks permit the inclusion of specifics in images of scenery as per the descriptions given by the users.

Development by Predicting

The unsupervised learning has the family of. The data is classified into a set of smaller predicted components. These models are taken into consideration to produce the effective results by forecasting the following section using these assumptions for input. One of the examples are the Language models, where each word is predicted on the base of the spoken word. Another example is the text predictor for the dictionary in a few messaging applications like WhatsApp.

The most striking difference in the text is the idea of unicorns in the form of "4-horned," Hence, the issues of understanding network relationships should be studied.

If the sequence of inputs is maintained , through which the predictions are managed, the autoregressive models are utilized to transform one direction into another.

The autoregressive models are utilized to discover information about data through the prediction of data bits in a predictable way. It is believed that a specific kind of unsupervised learning algorithm is designed using the assumption that any portion of data. It involves removing a word from the sentence and formulating it on the basis of the remainder in the phrase. The system is then trained to master the whole data through the method to make predictions. However the generative model could be misused. They allow the alteration of images and video for a lengthy period of time, so malware-related content can be removed quickly.

Re-Imagining Intelligence

The significance of models that generate data cannot be ignored. The ability of an agent to use the capability of developing information is an avenue of imaginationand is the possibility of making plans for the future. The research shows in studies that the ability to predict many aspects of the environment improves the agent's model of the world and improves the capacity to solve issues.

The results can be correlated with the wisdom of the human brain. The capability to gather data about the world with no an explicit direction is the primary motive behind the ability to think.

Implementing Supervised and Unsupervised learning in your company

It is the best method for developing practical business plans from huge amounts of data. Machine learning algorithms differ from one another on basis of how they go about using data to make decisions that is essential for various applications of business. Knowing how unsupervised and supervised techniques operate and the main distinctions between them is essential for the development of your company.

Supervised Learning

Supervised machine learning requires the researchers improve the algorithm in order to yield the correct result for a specific number of inputs. It generally involves to invest a large amount of data in the algorithm. It further defines the results machines are supposed to create using information input.

If there is a sufficient amount of data to train, accurate results will be reported by computer algorithms for machine learning.

It is a new kind of algorithm designed specifically for business , which includes the ability to train the algorithm in performing various tasks that range from adjustment of audits to in-depth data entry.

The supervised machine learning algorithms are used as cognitive initiation techniques to aid identification of pictures. They also conclude the knowledge gained from data that is not structured.

The biggest drawback with algorithm is the fact that it's involved in the extraction of data. The data needs to be manually arranged by humans prior to the instruction of the algorithm.

Data must be labeled conformity with the range of input variables, and then separated into a list of outputs that are possible. The more efficient algorithms will produce results when there is a sufficient amount of data used for training. This means that you need a massive amounts of historical data, as well as huge numbers of workers to support this process.

Unsupervised Learning

Unsupervised learning isn't as easy as supervised learning but it can open doors to specific applications. There is no initialization stage for unsupervised learning. The algorithm is provided with the data set and must use the variables already there in the dataset in order to identify and distinguish the natural clusters.

The advantage of this technique is in the fact that it doesn't need an intense arrangement of the data. Additionally unsupervised learning can be adept at identifying patterns that aren't visible for the human eye because of human errors.

The most impressive benefit of the unsupervised learning is segmentation of customers. In the same way it is agnostic of the biases that may be created when a company focuses on the demographics of its customers. It demonstrates that unsupervised learning results in the formation of segments of the customer that highlight marketing actions.

Different types of machine algorithms are utilized by data scientists, who can identify the patterns among a huge collection of data, resulting in useful insights.

Supervised machine learning is widely utilized by both. Methods such as logistic and linear regression are used and it also allows the machine using vectors. This is referred to as supervised learning since the data scientist directs in the training of algorithms as well as the results.

Supervised learning requires that the expected outcome of the algorithm is as well as the data that are required to provide precise responses.

For instance the classification algorithm can identify the pictures of vehicles when it is upgraded with a database of photos that are labeled with the vehicle's names and other prominent characteristics of recognition.

In contrast, unsupervised machine-learning is focused specifically on the artificial mind. The primary idea behind unsupervised learning method is that machines are able to independently discern the complex processes and patterns , even without the assistance of human beings.

Unsupervised learning may not be suitable for some business scenarios but it does provide the possibility of coming up with solutions for the problems that human employees would like to solve. One example of a machine learning is k-means clustering.

The supervised classifier is based on the already labeled images fed to it, while the unsupervised algorithm computes the common points among the models and differentiates the classes.

Chapter 5: Buildering Machines Be More Like Us

There has been a massive transformation in AI in the area of business , and technological innovations such as the automated doors and the ones seen in movies about robots have become part of daily life. In the present, we live in a world that the interface between humans and machines has been growing each second. In this kind of environment it's not surprising to anticipate more advanced AI advancements.

AI tools can aid in discerning between legitimate and fraudulent and spam content that is posted on social media platforms like messages, tweets, videos, Instagram posts, etc. The main issue are due to how computers can be not perfect and do not have the ability to read human language. They therefore consider any suspicious activity as be spam and malicious without considering it at all.

Even when the AI is set up to include typical errors and local phrases, AI still behaves as an imperfectionist. The participants are more social, more aware of context as well as more confident and unintentionally entertaining. The human character and the programming style of sarcasm tendencies are always a significant danger. It is evident that technological advances such as robots tackling hazardous tasks and automating complicated tasks are among the many benefits AI could provide to individuals of all types.

Martin Moran holds the director office at Inside Sales, a company which focuses on self-learning in sales automation technology. The company focuses on the use of software and customer services and administration as their top areas for research and development for AI.

He believes that AI is particularly useful for the crucial departments such as the administration department. AI is not than a matter of imagination and movies and is expected to prove its importance in the real world. It has already reached the stage of generation 2.0 of cognitive intelligence.

Moran believes that the next step in AI is likely to follow human beings more closely and mimic the human brain in relation to processing capabilities as well as access to huge data sets and the ability to interpret. Also, only when this AI know-how is embedded within the workflow system itself will the proper and long-lasting effect on business happen.

The possibility that we are able to interplay the complexity of natural language comprehension in the correct context with human behavior patterns takes us to a higher degree of AI system control. The capability to develop AI by utilizing real-world linguistic particularities will allow for real-world offices, businesses, and factories to employ AI. How can we create machines that look more than us?

It was stated that Splunk (operational intelligence company) that the power in AI can be observed in machines rather than the human studies.

The fundamentals of machine learning is the information that can be created by the data stored on these machines by humans analysis by machines.

Every human interaction with computers leaves behind a trail of information on the device. This information is extracted to provide us with a complete report of our specific human behaviour that ranges from how we shop online to the people we communicate with and where we travel through the device's geolocation.

A majority of businesses keep the majority of information only in small amounts but some of it isn't even tracked. When we begin to digitally capture and analyze our surroundings at an even more detailed scale, we can begin to create an AI that is more human like and reflects our actions more precisely.

The main aspect AI should be focusing on is the place where it integrates the most easily into what we can call the human interaction story. It is essential that AI information must be integrated into the structure of the company's work. Then, the brain of an AI begin to understand the world that is not perfect.

The world of today requires people to adjust to a world in which we must collaborate and communicate with machines every day.

A report released in the Project Literacy project, led and arranged by Pearson's educational services business estimates that the current pace of technological advancement facilitated through AI technology is enough to beat the rate of illiteracy of over 5% of the British adults over the next eight years.

Machine reading is not the complete understanding of the human speech. However, technological advancement means that robots are likely to reach the threshold for literacy in the coming decade exceeding the capacity of 16 million British citizens. The Harvard Amherst's Professor Brendan O'Connor.

The next step in AI could equip machines with the ability to understand emotions, just like human beings. In order to achieve this it is possible that the machine will receive data that displays different emotions. The data could come in the form images of sad, happy or angry people. The machine can understand the emotions by using these images. The data could also define the characteristics that are typical of every emotion, such as the setting and the language that are used.

Artificial Solutions worked hard to create a set of computer conversations which can serve to identify the nature of human behaviours and also to recognize our unpredictable conversational distinctions.

AI today is aiming to be as natural and sensitive as humans. This will require a degree of error when it comes to interpretation as well as recognition tasks, in order to imitate human behavior.

Chatbots Can Learn What You Are Looking For

The next stage of customer service is to be motivated by artificial intelligence, so that companies are able to concentrate on their customers in search of their preferred products.

In the past it was a bit naive to believe that you could utilize artificial intelligence to build an experience for customers that was more human. But, the AI nowadays operates in this way.

In order to make connections to us and develop interactions with our products that are insightful and personal, AI analyses a vast amount of psychological and behavioral data. The integration of Unilever's AI for the improvement of its tea brand , LIPTON's sales demonstrated the rapid adoption of AI in business.

AI can provide new methods to retailers to make sure that shopping is effortless. Cognitive engineering in the modern age, just like humans, can comprehend and reason, read and make connections. It's a rapidly evolving field.

To power their online shopping assistant IBM Watson, the company's cognitive computing platform, is operated through North Face, outdoor clothing brand. Artificial Intelligence helps shoppers complete their online purchases by selecting the appropriate jacks based on the answers they provide to questions about what and when they'll utilize it.

Macy's Macy's, the US department stores chain, created an algorithm-driven shopping smart phone application. The app allows customers to look up and ask questions regarding where to locate the item and also if the item is unavailable. AI is expected to continue being utilized to address customer issues and not to sell products. Advertisers can now invest in their advertising effectively. Retailers must be aware of trends in previous behavior and then make the appropriate adjustments to their the offers.

Today, the most skilled employees can master cognitive technology and apply this knowledge directly to the entire employees and customers.

Recent trends have been influenced several variables. For one there's been a rapid rise in the volume of data about consumer behavior. It is also more cost-effective and practical to make use of computers to aid in the development of AI. In turn, AI methods and machine learning algorithms are rapidly evolving.

AI has led to a significant change in the way companies consider the importance that AI when it comes to creating experiences for customers. Although many companies believed that AI was in opposition to human-centered experience design They now realize AI is helping to improve efforts. Although some companies believed that AI could have impacted interactions between humans but they are now seeing that AI is enhancing the experience.

AI is proving successful for customers selling clothes and cosmetics. A lot of online stores have implemented AI to fill the void left by customers who do not have the time as personal shoppers online.

There's no doubt that AI has enabled companies to provide a customized experience that predicts what will attract customers all over the world.

For instance, look-alike modeling this is when businesses can recognize current consumer preferences and then create an online platform to provide a similar set of new customers.

The use of similarity metrics can help to classify customers on the basis of their behavior when it comes to developing customer experience. AI has created a new revolution. The availability and proper utilization of high-quality data will enhance customer experience and consequent feedback.

The Mr. Singh who is a employee at the AI board level, suggests that current world is about competition with other people and the algorithms are the primary instrument and tool for the competition. The game is based on transparency, algorithms, and the application.

The next stage of AI is about flexibility which is where companies are looking to promote their products that are based on the individual's preferences.

For years, retail stores have been contemplating personalization for a long time however, it was only some status as an idea. Nowadays, personalization, or versatility is now possible for anyone with the capability to use AI in a mass. AI can customize interactions without needing to access the details of your personal data, just by looking for patterns in the online behavior.

Computers Communicating with Humans

Artificial Solutions Chief Strategy Officer Andy Peart says machine brains are being taught to comprehend and efficiently adopt human interaction by implementing Teneo Analytics Suite. Teneo Analytics Suite which helps the machine create sentences and arrange them in a certain order. Computers are becoming more sophisticated nowadays. It is crucial to remember in this regard that the abilities to recognize speech and respond to messages using AI is a complex procedure that cannot be compared to human communication and response. Nowadays, we expect machines to interact directly with us, comprehending us and contributing to giving us information about the subject being discussed.

What is the reason Computer Speech Recognition Bit Thin in the First Place?

Human languages are brimming with discussions, words, and a variety of

dialects and accents that differ. Furthermore, machines lack the ability to recognize homophones. These are words that have different meanings , but are pronounced the same like cite, site and sight. They all have identical sounds, yet have totally different meanings.

Why is it so difficult to make computers conversable?

The standard automatic speech recognition program has been released at no cost. It splits sentences into smaller phrases, and gives the ability to store conversational memories in relation to the context of the conversation. For example, it allows the virtual assistant to respond to conversations that have occurred before the user.

Conversation between humans on computers started with simple language for children. A few voices were developed specifically for children. The one I used was Speak & Spell included a handle and box as well as small green screen. It was used to determine an unfriendly tone.

Adults, the voice actors from the original cast were used for the computer-generated voices from in the nineties, eighties and the early laughs. If the announcement on the train announced that Port Chester was the next stop , using two words instead of Dorchester that is an indication of that it is a machine as a machine might not be able to comprehend that it was New York people used a single phrase instead of Port Chester. It is okay in the case of short phrases, but the machine language can't be used for lengthy sentences. especially in emergency situations the human voice is the only way to calm the fear and a voice of a machine that is void of emotion will only cause an increase in panic in such an emergency.

If we're planning to replace assistants by Google Assistant, or want an actual interaction with Alexa the device must communicate like a human. Alexa in this respect is required to react to signals from the spoken word and display the same rhythm and flow to human conversations. Computers should behave like humans and be beneficial to us. However, it's not as simple as it seems. The biggest challenge in this area is Prosody. It's the tone the accent, tension and rhythm , which is common characteristics of human voices. The issue isn't the words we're discussing, it's how we're speaking the words.

The melody is the main part of the human voice. Computers can be fed the words, but the process of teaching computers to communicate using all vocal expressions is a difficult task.

The most recent development by programmers have developed enhanced human voices , which are utilized by computers to communicate with humans.

It is a mix of four attributes that include frequency, tone, voice , and loudness. In speaking, people is able to combine many of these characteristics, however Siri is not able to accomplish this. The machine is restricted in one aspect or another, and can't mimic human behavior in a complete way. For instance, it can only talk about what it's been fed to it, whereas humans are free to choose various words and tones for expressing his various feelings of delight, shock and anger. However, we aren't able to input it into the computer. Programmers this can be a difficult issue since algorithms have constraints, however, there's no limit to our human voices.

Numerous tech companies have addressed this issue by choosing an authentic human voice with many individuals to be incorporated into their AI system, and then blending with them to create speech. The most recent developments by programmers have developed artificial human voices that are being utilized by computers to communicate with humans.

As voices become more powerful it is essential that the program doesn't mislead you. The person listening must be told that the voice is robotic and not human at the very least until the point that a fake voices are competing with real voices. Despite the difficulties that lie ahead, experts hope to create an audio model of computers in the near future. Achieving this goal will resolve various technological issues however, it will also create the most ethical and legal problems as it is possible.

Are Artificial Solutions Make Computer Conversations More Human-like?

It's not all about being human, however, we have to create an entirely new kind of casual realism which is quantifiable and could even be conversational, if you want. Human memories and interactions involving the meta-level knowledge of the world are feasible as long as the data is obtained through the Internet. For the vast majority of human interactions free-format , unstructured data can make it hard for computers discern the true intent of a user.

Chapter 6: Visualizing Predictive Model's

Analytic Result Results

in a sense, the outcomes from your analytics for predictive purposes need to be made clear to the most important individuals. Here are some methods of using modeling tools to communicate the results of your design to those who need to know.

Visualizing Hidden Groupings Within Your Data

The process of data clustering involves the method to uncover hidden connections between objects within your data. In the majority of cases data items that have the same kind of data are grouped together, including people who use Facebook, email, and text files. The graph illustrates one method to look at the impact of a data-clustering technology. The graph shows clusters that are found in the data gathered from social media users.

Data on customer interactions was recorded in a table. Later, the clustering algorithm was applied to the data . The three clusters identified were:

The customers who were wandering, the dedicated clients and discount customers. We can see that the two major components derived from the initial data are the two axes X and Y. The evaluation of the principal component is a method of reduction of data.

The interconnections among the 3 groups in this instance can also be explained by the importance and effectiveness of improved and targeted marketing campaigns.

Visualizing Data Classification Results

A classification system generally assigns a new data item which is being analyzed to a specific group. Individual courses could comprise groups that are created through clustering.

It is a moment of fact when you employ a clustering algorithm to classify groups of customer data The new customer and need the model to identify the characteristics of this new customer.

The image provides details on how to input data about new customers in the analytics predictive system. The system is then able to predict the customer group with whom the newly acquired customer is a part. In accordance with the model of classification the clients A, B as well as C, will be placed in clusters. The use of a classification technique will reveal the likelihood of customer A as a frequent customer, customer B being an unreliable customer, and customer C being an individual who shop when discounts are available.

How to determine if Your machine Learning Model Has the Correct Performance

What can the results of the model that was developed to make the predictions be analysed? This is a question that is frequently asked by students. The novices are always looking for an answer to this dilemma. An individual who is just beginning to learn wants to find out the proportion of predictions that have to turn into reality that the system is considered to be effective.

The next section will inform you about the assessment of effectiveness in your method.

Model Skills is Relative

The work in predictive modelling is distinct. It is comprised of individual information, the tools you're using , and the skills you'll be able to attain. The issue of predictive modeling hasn't been addressed previously. It's difficult to determine the quality of a model or what ability it may have.

Even if the model's creator is aware of the abilities needed by this model they might not know the extent to which it could accomplishment. The most effective thing you can do is to connect the machine learning models that are running on your data to other models that were trained using similar data.

Baseline Model Skill

Since the performance of machine learning is a matter of relative performance, creating solid baselines is crucial. A background to create forecasts on this predictive modeling task is simple and simple to follow. This particular model's capability is what defines a machine-learning model's worst output on the data set.

The conclusions derived by a baseline model demonstrates the place at which it is feasible to surpass the capability of other models that have been that are based on the same data.

Baseline model examples include:

* Calculate the mean for regression.

* Determine the mode to perform the purpose of classification.

* Predict the input input as output.

The baseline quality of your design could be used as a basis for comparing designs.

If a model shows an outcome that is lower than the baseline, it means that it's likely that something is wrong or the model isn't suitable to solve your issue.

What is the Best Score?

The ideal score is 100% accuracy if you're facing a classification problem. The correct scoring is 0.0 errors when working with a regression issue.

They are difficult to get. The entire process of predictive modeling could include a few errors.

The error can be traced to a variety of sources, such as:

* The noise in the data

* The sample of data isn't complete.

* Modeling algorithm's stochastic aspects.

It isn't possible to get the highest score, but this measure can help in deciding which measure provides a higher level of performance and lets you know the degree of variation that might exist between expected and actual results.

You can look up the possibilities of models based on your data and see the positive and negative outcomes.

Find Limits to Model Skills

When you've selected your baseline, you may opt to undertake an assessment of your predictive model and examine the quality of the model's outcomes. You must identify the most effective model, one that can be effective in making predictions for the particular data set you have.

The problem is not without solutions. There are two approaches you could explore:

* Begin High: Select an intricate and well-known machine learning method which is proven to work on a variety of predictive models like gradient boost. Examine the template that is applicable to your subject and then use the result as a benchmark. Then, look for the most basic version that can perform similar tasks.

• Comprehensive search: Review all the methods of machine learning you could think of for the issue, then select the best method for achieving the optimal base efficiency.

"Begin High" is an effective method "Begin High" approach is easy and will allow you to select the limits of capacity required to tackle the problem and then create a simple model that achieves the same results. It will allow you to determine if the issue is easy to forecast and is crucial since it is not always easy to predict the outcome of a task.

"Comprehensive Find" is a bit slow, and designed for long-running projects in which a good understanding of modeling is more important over all other concerns.

Both methods result in an overall score that can be compared with the baseline.

Why should you be sensitive In Using Predictive Models to score new datasets?

The process of creating a predictive model is complicated process that requires many assumptions, hurdles and an understanding of the real of what's being modelled. Models aren't just created by the impulse of an individual to build one. Models are constructed from a clear theoretical concept of the modeled reality. The designer of the model searches for the data sources which he believes could be utilized to construct predictive models for predicting the outcomes that are most likely to occur.

Thus, developing models that predict the same outcomes by using "good" historic data from the scenario being modelled is almost insignificant and redundant. Once a model is developed, it is able to be applied to new tables using the same historical information and contextual information used to build the model. The model is built with the understanding of the outcome being predicted and the historical character of the data changes.

For instance, an automobile insurance company employs a predictive model to approach potential customers who are interested to take part in their insurance policies however, they must rate the new customers to determine the risk of using various variables. Each client is rated by the insurance company must determine a rating that can be used to determine the amount of their insurance cost and premium, as well as the anticipated risk and risk.

In order to calculate the predictive scores, or the degree of risk that a potential customers who are new, the business could apply a previously developed predictive model to new customers. When evaluating the data set, it's essential that the table has all the variables that are used within the model. When the model is implemented with the new data table and a predictive risk score is calculated, which is used to determine the annual cost of insurance. The higher scores, you'll be lower the chance to be able to influence the insurance company, in terms of the amount of claims you make in the year. The lower you score is the less likely you'll make an application. It's possible to organize the scores during a downturn.

This method could have a variety of negatives. The hypothesis that is applied theoretically and in practice could result in negative results. It is crucial to be aware of the historical context for the data you wish to model or what you are looking for an appropriate model to be able to connect to.

Retail businesses can decide to build an predictive model to study promotional elements that have been proven to show superior performance. The historical nature of every promotion must be fully recognized. The company can get data on the success and failure of promotions by constructing a model using the historical data on circulation of a particular campaign and then applying it to different promotions.

Predictive Modeling Techniques

Define Predictive Analytics

Machine learning and mathematical algorithms methods can be utilized to calculate the likelihood of outcomes that could be expected from the prior information.

The focus is on going beyond descriptive statistics and describing what happened. Predictive models are backed by a solid results and can be utilized to generate predictive outcomes from new data. The results of modeling reflect the variables' potential.

Utilizing predictive analytics, many companies improve their processes and can gain an edge over competitors.

* The reason behind this is:

* It's quicker and can be used with less expensive machines.

* It requires the improvement of the quantity and quality of data as well as an enormous amount of attention to the production of important data.

* A complex economic situation and the necessity to stand out from the crowd.

Since the software has evolved to be more user-friendly and interactive, predictive analytics are not just available to statisticians and mathematicians. Similar methods are also employed by professionals in business. Modeling techniques that predict outcomes have evolved into the preferred tool of advertising experts to improve the performance of campaigns and to explain the investment return.

A systematic, logical strategy for getting the right information to the intended audience at the right time is accompanied by innovative methods to improve results. The capability to develop strategies to connect with a specific audience is more effective than ever if teams have access to clear information and reliable information.

This section will discuss some of the most common models for predictive modeling, with a focus on three distinct types:

* Propensity models

* Intelligent suggestions

* Models for segmentation

You'll be able to identify five methods of predictive modeling in each of the categories that your business could begin using immediately after you have the right tools to go about it.

Segmentation Models

The concept of segmenting the consumer in the field of advertising has been around for a long time. It was thought of in the form of an educated guess at one point but has evolved into an approach to strategic planning that is that is backed by data.

Here are some predictive modeling strategies which can help you create optimal audiences for your marketing campaigns, allowing you to maximize customer information.

* Behavioral Clustering

Leads are well-known to play a role to the conversion of customers, but the circumstances that lead to conversion are vital for marketers who are looking to leverage predictive analytics. The information known as behavioral data, is not important by itself in this regard. However, it is able to be combined with demographic data to help marketing teams to discover patterns and commonalities that can attract new groups of customers.

Through driving leads that fit the new target's background and demographics, as well as the same set actions, marketers can increase conversions as well as forecast the effect of their efforts in a timely manner.

* Estimation of Share of Wallet

This method explains the proportion of budget that is that is allocated for your service by the potential client. In the event that the customer has allocated a large part of his budget to the solution there is less chance of an increase in budgets through cross-selling and up-selling methods.

If this method is combined with clustering that is based on products to ensure that a proper proportion of the calculation of wallets shows the amount of money spent by customers against rivals, as well as the service and products you could provide to your customer to increase your wallet's share.

1. Probability of Churn

Most marketers solely to generate leads which isn't a great option as the security attributes of a model for predicting could be a fascinating application. Since it is known that keeping existing customers much easier than acquiring new customers so, every business should concentrate on keeping current customers. You evaluate the ability of a client to provide control by using the same process that is used to find leads based on actions. If the probability of a lead becoming to a new customer is correctly forecast, you are able to accurately assess the probability of a customer of going elsewhere.

Predictive Modeling is based on the Quality of Data

The availability of secure high-quality data that is that is backed by the top systems for data administration is crucial to all of this. High-quality data is the foundation of accurate predictive analytics.

Chapter 7: Neuralnetworks And Deep Learning

What exactly is Deep Learning?

Machine learning and deep learning are both components of artificial intelligence. deep learning is the more advanced version of ML.

Insofar as the subject of machine-learning is involved, learning process of learning from data involves using algorithms developed by humans who program. The information fed is used to make choices and take action. Deep learning allows to make use of an artificial neural system that functions like the human brain. This lets the machine understand data in the same way as the human brain.

Deep learning machines don't require any instructions or guidance from humans to complete their duties based on the data they receive.

Real World Examples of Deep Learning Models and AI

Concepts such as deep-learning and AI are not yet fully understood to the majority of us. A majority of people express mixed feelings when they encounter these terms first. What is the way machines can be programmed to perform tasks that are intended for humans? How can a whole industry that involves the development of machines that resemble human beings be explained?

These questions are important and require discussion. However they eliminate the uncertainty. If we're interested in exploring deeper learning or AI applications in the real world, we should read this chapter, which explains the ways that industries are changing with the help of AI as well as deep learning. Where did Deep Learning Originate?

Deep learning and machine learning are the two main elements of AI. The highest level of machine learning is deep learning. Human programmers develop algorithms that are able to learn from data and then perform analyses regarding ML.

Deep learning differs from machine learning as it operates with the basis of an ANN (artificial neural network) that is similar to our brain. ANN lets the computers conduct data analysis that is like the analysis of data performed by the brain of a human. These computers that have deep learning capabilities do not require follow human programmers' instructions.

Deep learning is a result of the huge amounts of data we generate and utilize on a daily basis. The data processing process in every deep-learning model is made possible by the data feed.

Deep Learning Simplified

The most dramatic advancement in human advancement is the development of artificial intelligence. If you've read this text, you need to be aware of the deep and ML. The book is now explaining the way deep learning functions and its place as a part of AI and machine learning. Deep learning is associated with human simulation and the technology.

If you're now well-versed in Machine Learning and AI You might be interested in knowing more what the purpose that deep learning plays in AI and machine learning.

What is the role for Deep Learning in AI and ML?

Deep and ML are both part of the AI class. AI is being used in a larger context. AI isn't a definite thing.

Everything that is possible by machines or software that replicates humans' intelligence are part of AI generally. It can be seen in a lot of contemporary technology.

So, in the sense of the following there are two types of deep and machine learning connected to AI. Both are interdependent.

In the same way deep learning can have specific properties associated with machine learning, but none of these specific properties are necessary to define the meaning and use of machine learning.

Machine learning is described as an AI branch that permits software programs to alter their algorithms with no human intervention. While traditional algorithms rely on engineers' involvement to alter their function machines learning algorithms are able to modify their algorithms based on data they are trained with.

When compared with machine learning, how is deep learning placed? As we mentioned previously, ML algorithms may adjust their algorithm to generate the right results. However, an expert has to manually alter the ML software in the case that its predictions prove to be not accurate. But, it is essential to label and arrange the data set that is used to train ML algorithm in a way that permits the program to grow.

For instance, you have for the algorithm to be fed with various dog and cat images to allow an ML-based program to learn how to differentiate between cat and dog images. It is feasible for ML to differentiate between two images if you label and categorize the data before feeding it into the ML algorithm. This highlights the difficulties in the application of ML. A sufficient amount of human-like intelligent can be produced by machines through the modification of algorithms according to the information supplied. The importance of deep learning in this phase is well-known. ML helps to activate human intelligence through mastering data and modifying algorithms. However the replication of functions of the human brain can only be achieved by deep learning.

Describing Deep Learning

Deep learning is one of the components of ML. Both ML and deep-learning are a part of AI and are built on the same idea of changing algorithms to achieve the best results without human intervention.

Deep learning is distinct from machine learning because it does not need to know how to achieve the outcomes it has predicted However, they can make use of raw data that is unlabeled and not structured to achieve the highest quality results.

Deep learning networks rely on the use of several layers of AI to create outcomes and process data.

What exactly are Artificial Neural Networks?

Artificial neural networks are designed to mimic the the human brain's neural networks. We will discover the way that neural networks that make up the brain to better understand the way ANN works.

If I ask you to determine what breed belongs to a particular dog from 10 images of dogs You will need examine all the images and ask yourself several questions regarding the dog's color and size in order to identify the breed of it.

The neural networks can answer these issues in fractions of seconds.

The same way, this scenario should be examined as well as accepted by an Artificial Neural Network (ANN) using different layers of the neural network. Each of these layers will employ their own unique sets of concepts and queries in order to tackle the problem. The end result will be to combine all the patterns that could be discovered through this neural network from the data it was fed.

So, this is how one can think about the operation of artificial neural networks.

The issue is what could ANN be considered to be an element of deep learning? Are they the two terms used to describe the identical thing?

Certainly An ANN isn't deep learning. In reality, it is many different ANNs.

Thus, Deep Learning is often referred as the deep neural system sometimes. But, the name is not a denial of their factual nature that they are integral to the use of AI like autonomous driving, among other applications.

NVIDIA carried out an analysis of an ANN called the generative adversarial system wherein the company presented the finest examples of the capabilities of the neural networks that follow. These networks were capable discovering patterns from the images to aid in finding new and intelligent faces.

The concept behind deep learning is now well-defined. We can now move to discussion on its use in the business world. Naturally these descriptions of deep learning that are perceived by laymen are only meant to aid in understanding the basic concepts associated with these ideas. The use of technical terms is best understood through a thorough research into deep learning.

The fundamental concepts that are associated with deep learning and neural networks are explained in the following sections:

More Power

Deep learning networks need high computing resources and typically perform GPU-intensive tasks.

Big Data Sets

The algorithmic development and layer development are not the main features of deep learning networks. Instead the primary feature that defines deep learning is the method of learning. Data is the basis of deep learning networks, and the quality of the data is a factor that determines the quality of content derived from it.

Very efficient Machine Learning Development Companies and Software Development Teams

It is not required for every company to invest their own money in the creation of deep-learning algorithms.

In actuality, engaging and having the most reputable IT businesses around the globe as partners is the most popular way to achieve an advantage in the market by implementing technology.

Deep Learning vs. Machine Learning

In the last few years, IT professionals are enticed by the new domains that are a part of Artificial Intelligence, which are deep learning and ML. This particular section can assist you in understanding the two terminologies in their most basic way. If you are able to continue reading this portion for a while and then the real difference between these two terms will be made clear to you, as well as how they can be used to create challenging situations.

The above album illustrates a lot of dogs and cats. Are you aware that you could get significant results if the reasoning behind these two concepts (deep learning and machine learning) can be applied to the pictures? It is evident in these photos that this is just an image collection of dogs and cats' pictures. If you're asked to select only the pictures of dogs and cats then you'll get a better understanding using ML algorithm and the deep-learning networks at the moment. Machine learning and deep learning: How machine learning is used to tackle this issue:

To allow the ML to recognize dogs and cats in the above pictures, they need to be presented together before the ML process. How will the algorithm determine the difference?

The answer lies within what ML is, i.e. it is based on data that has been structured. So, the images of cats and dogs will be labeled in a manner by which we can point specific characteristics of animals. The machine learning algorithm that needs to be taught is this information, and later it will work and separate millions of images of animals, and it will do this by using the labels it has learned.

If a problem is solved using Deep Learning?

To address this issue the alternative approach is being used by deep-learning networks. There is no need for the structured data in the images to differentiate between the two species this is the main benefit of deep-learning networks. Through the use of various layers, inputs are transmitted by the ANN network that employs deep learning.

In this process, certain features of images are systematically identified by the different networks.

It is essentially the same as problem-solving process that occurs in the human brain, in which questions are transmitted through a series of thoughts and related questions are asked to find the answer.

To categorize animals by their pictures, the system is able to find the right identifyrs after the data has been processed by layers of deep neural networks.

Note:

The above section was an illustration of two concepts that have fundamental differences. We also have described their operational and working areas. However machines learning as well as deep learning are not appropriate for every scenario in the same way they are both a good fit for certain scenarios, which we will discuss later.

In the same way, as we've already talked about that structured data is needed by machine learning algorithms in order to distinguish the cat's images from the dog's and to perform the classification in addition to producing output.

In the instance of deep learning images from both animals are classified by the processing of data in the network layers. The data that is labeled is not needed , since it is exposed to the outputs of every layer, which is how an integrated method of classifying images is created.

Key Tips

Data presentation reveals the major difference between deep learning and the ML. Data that is structured is required by the ML algorithms, while layers of ANN are required by deep learning network.

* In machine learning algorithms, the labeled data is required to perform the tasks in the machine learning algorithms, and subsequent outputs are produced using other data sets. However humans are not involved when the final output isn't what was expected to appear.

* Human involvement is not required in deep-learning networks, because data is processed by layers of distinct concepts and done through

layers with nested layers. In the end, learning happens through their mistakes.

It is believed to be the most important resource. The accuracy of the results is determined by the high accuracy of data.

What isn't Seen in the Example, Yet is Vital to Note?

* As structured data is required by computer algorithms that learn, they are not the right methods to tackle complex issues that require a large amounts of data.

Although we have witnessed the use of deep learning networks to tackle a minor task however, there is a greater dimension on which the actual applications of these networks could be seen. With regard to the sheer number of hierarchies as well as the methods they are not designed to solve the simplest issues. They are instead designed to handle complex calculations.

In reality, data is essential to both branches of AI to create any kind or "intelligence." Yet in comparison to traditional computer-based machine, deep learning demands an abundance of data which is the primary thing to understand. Because the differences are only apparent in the neural networks' layers when they are subjected to millions of data points. However, pre-programmed criteria are utilized in the ML algorithms.

In context of the above discussion and a variety of explanations of these concepts, we can now explain the differences between ML and deep learning terminology. We've focused on the basics of definitions, which don't contain a large amount of technical terms and difficult concepts for those who want to implement ML and AI for their business.

Below are nine Applications that make use of Deep Learning in Different Industries

Computer Vision

The deep-learning modules are widely used by top gamers. The cutting-edge detection of objects restoration, image and image segmentation are powered by deep neural network. Importantly, handwritten numbers on computers are identified by these networks. The strong neural network is the primary basis of deep learning, which is designed to recreate the human visual ability and is done through machines that are intelligent.

Sentiment-based News Aggregation

The deep learning algorithms are also utilized by news aggregators to blot off negative information and only reveal positive news events.

Automated Translations

Industry has witnessed the automation of translations prior to deep learning. But, computers are now able to provide better translations and with greater accuracy as a result of deep learning that was not possible before. Additionally, AI specialist can now find translations from images that were not achievable with traditional text-based interpretation.

Customer Experience

Machine learning is used by a number of businesses to enhance customer service. Self-service online networks are excellent examples. But, in order to create a solid workflow numerous companies are moving to deep learning. In this sense, chatbots are a common concept for the majority of us, which are utilized by many companies. Changes in this area will be anticipated because of the widespread application of deep-learning.

Coloring Illustrations

Much time has been spent in the production of media to enhance the color of basic black and white videos. But, amazing colors can be added to b/w images and videos thanks to AI or deep learning algorithms. As we are all aware, the color style is being applied to millions of black and white images to give them a new style and appearance.

Autonomous Vehicles

You will see a range of different models of AI working in concert as you see an autonomous vehicle that is passing smoothly. When pedestrians are highlighted by specific models, other models are recognized by the form of road signs. Many millions of AI models are able to train cars that are on the road. The AI-driven cars are thought to be safer than vehicles driven by humans by many.

Language Selection

With this in mind the above, it attempts to determine whether dialects can be differentiated through Deep Learning models. For instance the possibility of a person speaking in English is determined through machine learning. In turn, the differences will be revealed based on the dialect. After selecting the dialect and a separate AI that focuses on a particular language will be able to handle the extra processing. It is important to note that there isn't a single human innovation within any of the processes.

Generation of Text

From the beginning, text is now generated by computers. The style of the text is learned by the machines. The most valuable news items are highlighted. The variety of opinions pieces on human affairs is easily managed by AI specifically targeted text generation. Presently, articles on anything from academic issues to children's rhymes can be produced by methods of text generation.

Analyzing Images and the Generation of Caption

The ability to identify images, in addition to generating intelligent captions is among the best features of deep learning. In addition, due to the precision and accuracy of the image captions created by AI many websites are now recognizing the benefits of these methods to reduce time and costs.

Chapter 8: Operating Of Ai And Ml Projects Of

Companies

With a cutting-edge, powerful technology, such as the transformations made by ML, not noticing the hyperbole could be a tough decision. Certain, the ML projects are investing hundreds of billions. The base of digital transformation techniques is nothing else than machine learning. Naturally when people talk about machine-learning, they're either directly or indirectly referring to AI. It is therefore essential to understand the actual workings of ML that is currently in the practice of various businesses across the world.

Due to the AI capabilities, computer machines can now examine massive amounts of data that aim to reach an "reasoned" conclusion about the topic being discussed. This means that the decision-making process of humans is triggered with improved outcomes.

The application routinely used by AI is the main problem, even though it is simple to explain the ML and AI. The suggestions and matching of content of streaming content is one of the areas that has seen great success. As a result the online viewer experience is also being completely transformed. Instead of limiting human labor "expert" human effort required to sort, curate and break down the content into consumable types an analytical tool for personalized content delivery in the present time and age is none more than machine learning. Following the analysis of user behaviour and preferences the recommendations can be precisely tailored by various streaming platforms and the specific content can be pushed to the forefront to increase interaction and revenue.

In general, AI is required to be integrated by all industries into the business model of their respective industries. It's not required that the company's setup be massive to be able to offer AI enhanced services to provide improved customer service. Small and mid-sized enterprises can benefit from the advantages of AI. In addition, to facilitate the payment and collection process and inventory management systems, they need to be improved through prompt dispatch and delivery of high-quality products. Additionally, stocking errors and issues related to shipping should be taken into consideration.

Business domains where AI is changing the Landscape

Life Sciences and Pharmaceuticals

When there's a debate regarding death, a result is drawn by everyone that aging is a nebulous experience. Although you may not anticipate the length of time or endurance may realize that the your quality of life is destroyed because of injuries, joint pain and the exposure to diseases.

Yet, the aging process is able to be slowed down with the process of deep learning. Today, technology is utilized to identify biomarkers that are associated to ageing. After a while, body's parts that show wear and tear may be identified with a simple test of blood and, by applying medications and lifestyle modifications the doctor may be able to reverse the negative effects.

Food

Fresh fruits and vegetables account for about 40 percent of the grocery store's revenues. In this regard, the notion of a need to keep the quality of the product is akin to irony. Yet, giving statements are more straightforward than taking actions. The supply chains are largely influenced by the dynamic of the grocer. It can be a risky way to ensure that their inventory fresh and shelves filled.

In contrast, it has been discovered that the secret to fresher and more intelligent food is machine learning. In this sense, ML programs can be kept up-to-date with historical data as well as hours of operation and promotional data. Then, data scientists conduct analysis to figure out the quantity of each product that is required to be placed in order and then displayed. Furthermore, ML systems also work to gather and store information about weather forecasts or public holidays, as well as other important events. Then, after 24-hours, the suggested order is issued to allow retailers to ensure they have their inventory of products.

The rate of out-of-stock is controlled at up to 80 percent with the companies that have implemented ML in their workflows. In addition, a the 9 percent increase is evident as a result of gross profit.

and Entertainment. and Entertainment

Due to the ML technology technology, the content from media houses is now accessible. Deaf Americans and people with hearing impairments are now able to watch YouTube videos by using the automated captioning software based on ML technology.

Information Technology

Many business perceptions are derived from machine learning, AI technology is not utilized by a lot of companies. But, by 2020, there will be 2.7 million jobs in data science.

Law

Deep learning software plays crucial roles in the legal field. It is often difficult to understand legal terminology but more than 10000 documents are analyzed by deep learning programs.

Legal professionals used to have to look through the stacks of paperwork manually in order to look over the contract clauses that affect the business of their clients. Today, contract clauses are able to be integrated by a computer program, that will respond faster and will identify key phrases for further analysis.

Insurance

Everyone tries to reduce the risk by countersigning and certifying details. Thus the insurance industry could benefit greatly from ML. To assess the amount of risk, data from customers and real-time information can be used by machines learning algorithm.

Based on information, rates may be adjusted by algorithms, which is how savings are generated for both insurance companies and consumers.

By conducting a thorough analysis we can alter this process by which, isolated social media data is gathered by ML software to construct an authentic profile. The individuals who appear to be healthy can be identified through the insurer with AI.

The person who is responsible for those areas of their lives will be mostly an accountable driver, too.

Education

Students can now be taught via Intelligent Tutoring Systems (ITS). In this case, the AI platforms function as virtual tutors. Depending on the strengths of each child, their lessons are adapted accordingly. The information is analysed using an ML program that can personalize the future content, each time a quiz is completed by students.

Furthermore further, Furthermore, the Intelligent Tutoring System ensures that the student has a wealth of knowledge and can overcome problems with learning. This is achieved through "learning" the specific requirements of each user and deciding on the kind of instruction that is beneficial to students. According to the results of a study students who are using intelligent tutoring programs are highest performers compared to those who are taught in a group.

Health Care

The cost of health care per individual by the US is reported to be higher than other countries. For instance, about ($3,749) for each person annually is paid for health care in the UK and is lower than those in the United States. It's very unfortunate that health outcomes aren't as good for the USA regardless of the huge expenditure.

In in the United States, the health cost of healthcare can be reduced by the advent of AI which means that the number of tests can be cut down and the correct decisions can be taken with life-saving outcomes. The potential for exponential growth is likely to be seen in the implementation of AI due to the large costs of health care and the benefits resulting from healthcare decisions.

In what ways AI and ML are enhancing the Experience of Customers?

What can the customer experience be enhanced by the use of machine learning and artificial intelligence?

A significant connection between online shopping was demonstrated by ML as well as AI. Without recommendations, Amazon or other shopping platform cannot be used. Based on your personal characteristics, for example the history of your purchases or browsing history, and more the recommendations you receive are usually taken to be personalized. The mentioned online platforms will likely provide a digital representation of the salesperson who recognizes your preferences and may provide you with products that you would like to purchase.

Everything begins with quality Data

A bit of heavy lifting has to be done at the back end in order in order to make this decision. Who are your actual clients? Are you aware of them? A collection of data and their track history is left by nearly every customer. However the data comes composed of fragments. Connecting these fragments to one another is not an easy task. If there are several accounts belonging to a customer how do you go about to find them?

If a client has different accounts for professional and personal activities are they able to connect the two accounts? If multiple names are utilized by a firm What is the best way to identify the only company responsible for these accounts? The customer experience is about the strong connection you have with your clients, as well as knowing the relationship between them. The idea behind entity resolution is to eliminate duplicate entry in the customer's list. It is used by large-scale companies that manage the massive data teams. In the present, it is a democratization of entity resolution that is taking place. Due to the absence of startup companies, small to medium-sized companies are not able to access resolution of entity services.

Once you've identified your clients the next step is to build an excellent relationship with them. The first step to understanding the requirements of your customers is to gain a broad overview of their behavior. What type of information do you've got about them and how it is going to be used? The process of acquiring data is to apply the technology such as AI as well as ML. This can be a tense process and controversial in the context of processing data streams from various apps, sponsors as well as other sources. When you are establishing data about your customers, please ensure you get their permission and ensure that their personal data isn't compromised.

ML is very much like any other type of computing. It is possible to still apply the principle of "garbage in garbage out" in the ML domain. The outcomes will be lowered due to poor-quality training data. Due to the increasing number of sources of data, there is an increase in the amount of data fields and variables. There is also an increased chance of errors for example, typographic errors, transcription errors , and so on. Manually fixing and repairing data is usually possible but it can be difficult and prone to errors to fix data manually for the majority of data researchers. The research topics are currently repairs to data and data quality and data quality, both of which are designed for an entity solution. In addition to help with automated cleaning of data, a new generation of machine learning tools are emerging in the present modern day and time and.

The most common field is recommendations systems, personalization and systems in which the uses to ML as well as AI are discussed regarding customer experience.

Modern times are witnessing the most well-known tools that include the hybrid recommender systems applications that mix multiple recommendations. Different sources are needed for most hybrid recommendation systems and they are paired with deep learning models and huge quantities of data. Personalization technology and advanced recommendation will be live and the recommendations should be based on models that are only adjusted periodically. The systems that train models against real-time data is possible through online learning, reinforcement learning and bandit algorithm.

In addition to improving customer interactions as well as improving customer service, many work processes and business tasks are being improved through algorithmic learning or AI model-based models. Chatbots are among the best examples of this. As of now chatbots have no actual use. But, they could result impressive rates of customer acquisition provided they are developed efficiently. However, we are still just beginning to understand the basics of natural processing of language and its interpretation. Several developments have occurred recently. Due to the changes in the creation of complicated model languages, Chatbots are now transforming from sending notifications to handling simple questions and answer situations.

This revolutionary shift is expected to change the way we interact with each other. chatbots are likely to prove to be an integral component in human service delivery efficiently and smartly method. Personalization and real-time recommendations is expected to be integrated into chatbots to provide this level of efficiency. The bots must discern clearly between humans and the customers.

The machine learning technology is being used in fields of fraud detection. Due to the rapid growth of security concerns, there is an ongoing struggle between criminals and the legitimate people regarding fraud detection. The most sophisticated methods for online crime are used by criminals. The fraud that involves people-to-person is not a thing of the past but it's now designed to mimic the bot that buys all tickets for an event, meaning that sellers could then resell the tickets. As we have seen in the recent elections hackers are able to gain easily accessible social network via bots, in which chats are filled with automated responses. Finding these bots and blocking them in real-time is an arduous task. Machine learning is able to accomplish this task. However, even then it won't be an easy task. However, the solution is equivalent to re-designing an online community, one in which people feel loved and safe.

In the case of automated customer interactions, the amount of friction could be decreased by advancements in speech technology and detection of emotion. Being able to respond to customers in a suitable way will be easy thanks to multi-modal models in which different inputs are integrated. In turn, customers can now express their wishes, which includes streams of live video.

When people think they are in a "mysterious valley" due to the interactions between robots and humans however, robots will make future customers feel more comfortable more than what we currently have.

If the an uncanny valley can be used as a way to study the customers the value of their purchase, then acknowledging their worth is essential. AI or ML applications that interact with customers will need to keep their privacy and the apps must be unbiased and secured. They could be complicated to some degree; however the customer experience won't be improved if customers are to feel ignored. There is a more effective solution, but it's not the ideal solution.

What can the experience of customers improve through machine learning and artificial intelligence? Many new developments are yet to come out due to these new technologies. Additionally, AI will now be utilized to create a seamless customer experience.

Chapter 9: Step By-Step Method To Develop Ai

And Ml Projects To Help Business

The AI and the tech sector are interconnected. In this way, we have witnessed a variety of developments from automated customer support to high-end data science. AI is getting worldwide fame nowadays.

One of the most powerful forces in the field of technology is artificial intelligence. It's also an emerging issue at conferences, and is engaging industries like manufacturing and retail with its astonishing findings. Virtual assistants are advertising new products and services as well as answering customer questions. are answered by chatbots.

AI is being integrated in the form of an intelligent layer for large corporations like Google, Microsoft and Salesforce in their technology systems. Absolutely, AI will conquer many commercial and business areas.

It is introduced gradually into a variety of commercial industries including: Skynet, Music industry and intelligent graphic user interfaces. The AI is astonishingly placing its mark on almost every aspect and as a result of the fact that our technology is becoming more sophisticated and allowing for the creation of big data as well as revealing new methods of data acquisition. This is a sign of enormous advancement within deep learning, machine-learning, and natural learning. Therefore, the process of integrating an algorithm in cloud platforms is now streamlined and simplified.

When it comes to businesses involved, AI can reveal diverse useful models that are based on business intelligence that is derived from data as well as your needs for your business. For businesses, AI can perform extraordinary tasks, i.e., from gathering data to effectively managing the relationship with customers. It also makes the most efficient use of logistics by effectively monitoring and maintaining different resources and assets.

ML is also a major factor to the advancement of AI. In the present, the developments that are taking place in ML are driving AI models. It is not enough to focus on the one thing that has worked and the economic benefits that can be derived from ML are far beyond what we expected at present.

From the perspective of an enterprise, there are some crucial business procedures related to coordination and control could be affected by the current events. Additionally, these processes can also effectively organize the reporting and resources.

This section will provide the latest information and tips on the steps to take in integrating AI within the company, in addition to the need to ensure its success

Be familiar with AI

We must be aware of current AI and the awe-inspiring results. It is the TechCode Accelerator with diversified resources is a great resource to use through coordination, such as: initiative from Stanford University. University of Stanford. The resources online can be accessed to find out the basics of AI. The platforms, like audacity and code academy , are able to provide the online courses. From there, you can start your journey.

Below are some additional online sources:

* Online lectures offered by Stanford University.

* Open-source Cognitive Toolkit from Microsoft

*Open-source library of software developed by Google.

Click here for the research and development of Artificial Intelligence.

Choose the Problems that you require AI to solve

It is the next stage to investigate the new ideas once you have acquainted with the basic concepts. You could think about incorporating the AI capabilities into your existing products and services products. The important issue is to remember and use particular models for AI applications that address the ever-changing challenges faced by businesses.

In the course of executing your duties in the business world you should be able to summarize its primary tech programs and issues. This will include the compatibility in integrating natural language processing also known as ML and image recognition to the items and products. For instance when video surveillance is carried out by the company the company, it could improve its value by incorporating ML into its workflow.

Prioritize Value-Added

The financial costs of the company will be evaluated during this phase across various AI applications. The parameters of feasibility must be reviewed in order to determine the priorities. This provides you to make a decision on priorities and determine the value of your business. Acceptance and ownership would also be sought after from older employees.

Accept the Internal Capability Gap

The goals you set for yourself and the ones of your business differ over a certain time frame. So, before moving to AI businesses must be aware of the limitations from a technological perspective.

It is sometimes tedious. However by focusing on your strengths and strengths, you can determine the areas you need to complete.

Many teams or projects can aid you in reaching this goal based on your existing company.

Contact professionals and create a your Pilot Project

after establishing the business following the establishment of the business, the next step is to incorporate it with an AI model. The trick to success in this process is to set smaller and manageable goals, while taking into consideration the guidelines and rules of AI. External experts can be brought in at this point.

Don't spend too much time on your initial page. In order to carry out a pilot project in a short time frame, a duration of 2 to 3 months may be enough. Following the completion of the pilot project, you must consider your long-term plans. Additionally, it must be determined if the value proposition is important to your business as well as AI experts.

Start Team for Integrating Data

Prior to integrating ML to your company Cleaning the data available is the first thing you need to do. Data storage towers contain corporate data that could be shared with businesses with a different goal. To counter this the need to establish an organization that can clean the data and get rid of any anomalies is the best method to ensure the collection of valuable data.

Implement Small

The process is properly managed if the work is performed in small pieces. Alongside carefully reviewing your feedback carefully, smaller steps can be executed to get desired results. The feedback will slowly lead to improvement in your work

Do you have storage plans within your AI Program

In order to integrate an AI solution, storage requirements have to be considered following the shift from a smaller amount of data. It is possible to achieve excellent results by improving the algorithms. But, to create accurate models require a lot of data. Additionally, the computing requirements are not achievable with AI technologies without the availability of massive amounts of data.

Furthermore, AI storage for data modeling and workflows must be enhanced. This could have a profound impact on the functioning of the online system, provided that all options are explored.

Include AI in Your Daily Plans

The advantages of automation and insights is provided by AI models, and employees who work in the IT sector can benefit from the benefits of incorporating the AI model into their everyday routine.

Nowadays, workflow-related issues can be resolved with the aid of technology. Corporate organisations are a great help in this area. Employees will also get a chance to learn about new concepts from AI as well as experience a shift in their work.

Include a Balance

The demands of the technology and research project must be met when designing the AI system. The creation of equilibrium in the system is the most important factor to think about prior to designing any AI system. This is a fact however, in a lot of instances, specific features are included in AI platforms in order to help reaching research goals. A few important elements are overlooked, for instance the requirements for hardware and software that enhance the research could not be considered.

To get over these limitations to overcome them, businesses must provide adequate bandwidth, network space and bandwidth for storage and an efficient graphic processing.

However there are other elements to consider, such as the provision of budgets to ensure the power supply in case of power loss should also be taken consideration.

How to Create an AI Startup?

Presently, AI is gaining widespread popularity. Everybody has been talking about the benefits of artificial intelligence, and the task of avoiding this hype is difficult. Without getting into in-depth knowledge and confusing subjects this article will help you aware of the fundamentals of AI. Then, you'll be able to identify the four basic steps that are needed when creating your AI business.

It is equally important to take into account that, with the increasing attention paid to AI it is possible to believe that they is far behind even if they do not make use of AI. AI platform. But, AI in many forms is not yet advanced enough. In addition, it's hard to determine the level of level of complexity. It is also unclear when is the best time to initiate AI and when it's not the best time to start it or not.

The prevention of AI is not the case when you're willing to start a new program. In the event of numerous issues AI technology can help. AI technology can help in the process.

AI at the Conceptual Level

It is not a crime when someone is unable to grasp the significance of AI. A lot of people think it is their smartphone or robot at their home? It is often difficult to understand AI after examining its artifacts. Furthermore, debates over AI and. ML are on rise. However, despite the differences, the broad issues will be discussed on in this section and we will be able to see their connection to the challenges that entrepreneurs face.

However, AI is a software application, in which input is taken in and output is displayed, similar to other software programs.

The main difference is that a simple step by step instruction isn't provided from the AI program to complete an alteration, and AI program could be unaware of these steps.

There are many ways to make the AI as a software application. It could be integrated into an application, a voice-powered web or any other device.

If the AI is discussed in this context, we simply are expressing the software component that receives input and generates output. The reason is that it takes only a few seconds to change the input into output, in which case, writing precise commands for the process of transforming input to output is difficult.

Humans can complete in a better manner than computers. Humans don't always deliver the performance that is desired in everything.

Common scenarios include the highlighting of sentiments from the text, identifying an image, or by looking for subtitles, or scanning the results of an examination for medical reasons. But, significant deviations in the results that were anticipated.

Classifying AIs

AI is able to be divided into two types: first one, where normal tasks are executed in various ways like transforming spoken words into written words. Then the second kind one, in which the system is able to handle tasks that are of a variable nature such as whether or not a heart issue is detected by an array of heartbeat data. The shift is essential and decisive because the intelligent systems have already tackled the most common issues and the available AI can be utilized rather than creating your own.

AIs that are already in use

The AI offers the highest performance when solving various issues. Face detection, or recognition of speech in photographs could be instances. Because of their nature issues developers have created a large portion of the AI applications based on AI to tackle these issues. The difficult part is already created and we can benefit from these applications.

For various purposes, many AI products have been created by the cloud giant, that can be used on models, such as: "pay as you use".

In addition, as they can be AI powering services, this wouldn't be relevant if you didn't make use of these services. You must be aware the possibility that these services will give you the correct responses to your inquiries.

Before creating your own AI make sure to search for an existing and published AI that you can use, to avoid having to rework it.

Custom or bespoke AIs

Custom-built incredible feature is the next level of AI. You could create your own application, when you have to complete something that's unique enough for existing technology to be accessible.

It's essentially an easy task, however it becomes difficult when you get to the finer details.

123

A Brief Overview of AI Information

Many different techniques are out there, but the neural networks are those that are getting a lot of interest. It is basically an network of neurons that are connected to each other. The system transmits a signal to the initial neuron that may or not connect with the other neurons. A signal output is created at the other end that of the. For instance, a list that contains the faces in a photograph might be included by the output signal.

The development of a neural net is possible in two ways: building the network and teaching it. The number of neural cells and the kind that they connect to can be determined prior to the construction of the network.

It is necessary to build the neural network following creating it. This basically means that mathematical functions can be applied to each node, and this function will inform the node of when to transmit an input signal and when it is not. Fortunately, this can't be performed manually, as it appears to be utterly absurd.

In order to guide a neural system the framework for training is used to give an overwhelming amount of information to the system. For every neuron, the math functions are created within this framework.

The synergy of these two functions, size and the linking of neurons is called an model.

Once you've created a model many containers are available that can be loaded into the model. Additionally, you can upload your model onto an online platform using the certain standards.

When a person is working on an AI to integrate it into the corporate environment and the kind of information that he or she will require in addition to knowing the source of it will be the biggest challenge.

Recommendation to Person Creating AI-based Startup

This involves four steps:

Find the Problem and Solution that is Right for You.

Your venture will not be considered legitimate and functional in the event that you do not address an issue that customers are willing to pay the cost.

Before you take any big decisions there are a few aspects that need be considered for example, identifying the individuals and ensuring that you have enough money to fund the work you're planning to do. Furthermore, you can verify whether the project you plan to perform is feasible.

When you implement solutions based on an old-fashioned lean approach it is important to consider whether your solution will entice the public and if they are willing to buy and accept it. Making a basic version of your solution is simple and involves using real people and other components. This is one of the best things about AI.

AI services as well as existing AI services and other humans who are involved in important events can work together to create the prototype that the solution you envision is simulated. In turn, you'll be able to conduct test solutions for your product before constructing a broad-based AI.

There is a question to ask: why AI is required to be developed? Are you able to determine if it is a smart solution or a solution to the issue? Each situation is different and it's not required to assume that every issue can be solved with an established solution. But, if real humans are being used to analyze your issue, take an instant and ask the following question: "What is there in my situation that requires an AI could be the perfect solution?"

It's true that it will not be an easy task however it is important to determine whether or not the software developers have the ability to create the AI upon which you're trusting. Making an AI isn't the solution to all problems. Making this on yourself could be difficult. However, it is possible to get the assistance from an expert who will assistance in reaching the milestones at various stages.

AI-Building Game

It is possible to advance further once you've realized that you've built relationships with your customers and you believe that your AI could be developed today. So, the first version of the AI could be born. As we have already mentioned the process involves acquiring some information, then purify it and verify its value. The next step is creating models, and then training it.

In light of this it is important to be aware that the most crucial and difficult aspect of the issue is the process of finding data, managing and coordinating it. A majority of the time spent by computers is spent in the process of creating models however, human insight remains essential in two areas, i.e., acquisition and comprehension of information, and you will have to devote the majority of your time in this stage.

Create Your Product

A functional and working AI is in place in the present moment, However, the users may initially have difficulties getting tasks done. As of now, you're aware of how to train the models, yet your users have no knowledge of the process.

Perhapsthey'd like to launch an app on their smartphones without having a clue about the technical aspects or converse with their home assistant that is voice activated.

To accomplish this for this purpose, your AI must be packaged into a service with a user interface the midst of other things.

Remember that a product can be considered impressive in the event that it provides solutions to the real-world problems. The existence of an AI program that analyzes images and pinpoints the exact position of faces isn't significant. If we can identify names from a set of images, something in advance of this is required.

For instance, that AI must be packaged in a package that shows the original image of the user, and containing boxes around the person's faces. Furthermore, the AI might require the user to write the names of the people who are who are in the boxes. This way the idea of flash card will be created and their function is to remember the names of everybody whenever they see this image in the future.

Create Strategies to Enhance Your AI

Your AI can turn out to be effective provided it is able to be trained using good data. When your startup is functioning mode, you will encounter a large amount of data. In addition, you'll be able to see the data that was absent at the start of creating the AI.

In the event of collecting data, you'll need it to train your employees. The most crucial thing to be thinking about is what method the data will be evaluated to determine if the performance of the new generation is higher than the previous generation.

In conclusion this chapter, it's just an introduction to AI to motivate you to attain your goal. Moving forward is to begin making and testing the personal AI or other. If that's it the case, it's a great opportunity to inquire.

In any case, interacting to your team of technical experts should be the next step. If you do not have a team, you may seek advice from a senior person to get more details and clarifications.

Additionally, you could help you in responding to your queriesif have experience developing projects, are knowledgeable about data sets, and a knowledge of the practical application of AI.

Chapter 10: The Future Of Ai

While organizations attempt to handle a variety of information and a variety of gadgets, nevertheless high-quality decisions can be made using AI or the IoT because of a couple of brand new methods.

Jane is a specialist in marketing. She collaborates with her group to create a proposal to a prospective business. This client happens to be a massive business conglomerate and the team that is presenting is under pressure to win the client.

Additionally, Jane data collection process is also complex. Jane gets data from colleagues, who are remote and transmits data via email as well as phone calls through WhatsApp as well as instant message. She is also seeking out information on the internet to ensure that she has completed of her task.

In 2020, around 7.6 billion people will constitute the global population. In the same year IoT connected devices are expected to reach a staggering number of billions. IoT gadgets that are connected to the internet are predicted to increase from 20-30 billion. In this situation how do Jane and the tens of thousands of others manage this amount of information and determine what is important, besides making the right decisions under these conditions?

If these tools as well as human resource are worked out, we will see an increase in data. This abundance of data is already causing "infoxication.'

Artificial intelligence appears to be one of the most effective solution to all these problems, and certain tools are created that are capable of dealing with the flow of information, as well as search for reliable sources of data. Furthermore, informed decision making and better cognition are also advantages.

Workplace hub is generally focused on the office and in particular, the coming workplaces. It is one central platform that brings together all of the organization's technological capabilities.

Additionally, it improves efficiency of the company by reducing IT associated expenses.

It provides real-time information by which business processes are transformed. With the advent of IoT and AI technology, workplace hubs will change in the coming years and will become a hub of cognitive thought.

Intelligent cutting-edge computing technology and AI will be integrated through this latest technology. Alongside enhancing interaction between groups and individuals humans, their intelligence will be enhanced, allowing for the expansion of the human interface network.

In the digital space, Cognitive Hub will serve as a flexible platform to exchange information, and provide enhanced intelligence for the masses.

Additionally, it will incorporate the latest gadgets, like flexible screens and augmented reality glasses as well as smart-walls. In order to incorporate intelligence, Cognitive Hub will utilize AI to process and collect information with the aim of create a sense of comfort for employees, teams, and individuals.

According to certain people, cloud computing will disappear however the reality is that it will continue. Instead, it will change and evolve into an artificial cortex that is a intricate three-dimensional trees. Cloud computing today is a link between cognitive computing and intelligent automation, as well as other AI driven areas.

A lot of work remains to accomplish on the cognitive hub, but, our work style has already changed thanks to the workplace hubs and allows us to manage the increasing intensity of technology and information.

The future AI in 2020 AI in 2020

In the context of Millenials and the generations that will follow Modern-day advancements differ from our past. As of now, our minds have created simulations of virtually every aspect. Brain was , and is the only commonality in our lives, including our ancestors as well as the coming generation, which will change our ways of thinking, communication and patterns of work. Researchers have been predicting Artificial Intelligence since long. It was, however, initially associated with robots alone. Nowadays, AI has been integrated into virtually every aspect of our usage. AI is generally regarded as a software that pretends to behave as humans.

It's a great benefit in the present day and agebecause it has brought comfort and peace to the lives of people and allows us to enjoy the speedy completion of work. So, only minimal energy is used as well as time and work is completed quickly.

Programming

It has led us to a the realm of new wonders where the world is evolving beyond our imagination. AI is the source of connectivity and provides the continuity between distinct actions. We could say that it is possible to communicate with a variety of things at a time for example, a quick conversion between languages. In the early days of computers, we've followed certain protocols to manage our actions. These procedures can be altered from the settings menu. These rules do not have to be followed when transferring the same technology to AI. The algorithm is instead taught to recognize the sequence of actions.

Predictions for 2019

We are exposed to a range of variables within Artificial intelligence. All variables can be processed using a programming way, which is simple and provides an obvious understanding and trust in the transparency of nature. It's a debate, but predictions of future can be reliable in the realm of data science. Since the beginningof time, computers were thought to be able to handle all of the tasks of statistical and numerical. The basis for machine learning lies in the computation of statistical significance and plays an important part in forecasting conditions in the environment, diagnosing ailments and playing the game of chess. We are constantly moving to greater processing power and huge amounts of data and the computational tasks facilitated by algorithms are the best.

The Decisions

Based on the data gathered through management data systems most decisions currently are taken by companiessince the data is directly derived from the company's operations, as a consequently it's hard to determine the right or wrong rules. Decision-making gets honed with the integration of AI in decision support tools. As they transform the data of customers as predictive models decision-making tools are also supported through the proficiency of AI. Based on important demographics, departments, like marketing and consumer services are altering their work due to this trend.

It is a new technology that is which is supported by digital banking and other enterprise applications currently use this technology.

Interactions

With minimal effort, new designs of the interface are developed through the AI. The advent of mice and keyboards has allowed us to make use of these devices throughout our lives, and we're still users of the keyboard and mouse. In the area of digital communication is concerned we've been able to develop and design algorithms that allow for the benefits of realizing. We can now observe an effortless and natural interaction with the code, which are now transformed into human language and can mute the input from camera and sensor.

What are the things you should expect from Future of AI Technology?

Our thoughts, actions and ways of living are changed by a variety of technological advances, but the dramatic changes are brought about by AI. Although AI is just beginning to enter the market, AI has become more flexible due to the technological advancements. When evaluating its future, a person is able to observe an environment in which every aspect of our lives is controlled by AI.

The Most Promising AI Innovations

In general, our entire lives will be transformed by AI acceptance. Whilewe search for AI -driven tools for our home however, corporations, businesses and even administrations are employing AI.

The introduction of autonomous cars on the roads is a prime illustration. While the industry is predicted to expand the regulations and guidelines to regulate AI-driven cars are being developed by the U.S transportation department.

AI in the area of transportation seeks to create self-driving vehicles. In the present, AI has achieved the characteristic of creating automated cars that are driven by humans and is set to achieve autonomous vehicles that do not require human involvement.

To create self-driving planes as well as busses, AI remains the main focus of many companies and the transport industry.

AI and Robotics will Integrate

AI has been coupled with cybernetics, which is a continuous development. Through the integration of AI technologies into robots humans can improve their bodies through power and endurance. While we could enhance our bodies through the technological advances, disabled could benefit from the benefits. The lives of those with amputated limbs , or permanent paralysis is enhanced.

AI will develop fully functional Robots

From Nexus, to Above, we are able to develop artificial life-forms due to AI technology. Human-like robots that are capable of performing intricate interactions has been widely studied by science fiction writers. Robots have a variety of importance as the field of robotics is changing due to the presence of AI. For instance, robots could carry out a dangerous job and perform actions that isn't safe for humans.

Impact on Humans

Gartner has released figures showing in which 1.8 million jobs could be eliminated by AI by replacing 2.3 million jobs until 2012. In analyzing this process, which began with the 3rd industrial revolution and progressing to the present digital revolution and observing the fact that we have seen an enormous shift in our lifestyles as well as our working habits and that will continue to happen in the coming years. Imagine an era in which people work just two days per week. This is the future anticipated.

But, AI is a big-big deal when it comes to the synthetic elements. We are living in reality, as with the advancement of learning, we are accustomed with the speedy accomplishment of tasks with the highest accuracy.

Automated reasoning is positive employment, 2020 is likely to witness the development of businesses in AI.

The figure of nearly two million net-new services is predicted to be reached by AI related jobs until 2025.

An unfortunate employment incident that happened in the past is often linked to technological advancements. However, quickly it will be replaced from the advantages provided by these advances. The company will evolve in the near future and the direction will likely be followed by AI. The efficiency of many job titles will be improved by AI and will eliminate various low-level and center-level posts.

The Future of AI in the Workplace

AI and IOT are not just creating smart homes. They are also affecting many businesses, in addition to interfering with the workplace. The efficiency, productivity and precision of a company can be improved by AI.

Yet, many people fear that because of AI advances, robots will replace human workers, and are viewed as a threat rather than being a tool that can transform us.

As the debates continue on AI in the year the year 2018, it must be acknowledged by businesses that self-learning and black-box possibilities is not the way to go in this day and age. With the realization of AI results based on data to show the value of adding value and to increase our human capabilities, the limitless potential in AI is being felt by numerous companies.

A number of decision makers have started to utilize the power of AI due to the benefits of AI systems are confirmed by numerous evidences. According to research conducted by EY companies that are integrating AI at an enterprise level are able to see two primary advantages: efficiency improvement as well as informed decision making.

The competitive advantage is earned by the business that is the first to implement AI. It can reduce regular expenses and could decrease other headcounts. This is a good thing from a business perspective however, individuals in various jobs are most likely to be replaced with machines. This isn't an ideal thing. There will be some tension between humans and machines is likely to arise with the advent of AI.

With the advancement of these technologies our economy will be impacted by AI through the creation of skilled jobs.

It is expected that AI will come into the picture to replace specific jobs that involve repetition, and the human capabilities currently in use will be hidden. Humans' place in the world will be determined by the AI tools.

The tasks, for example the detection of fraud, loan approval and financial crime , will be carried out by automated system.

With the progress in automation, an increase in the production efficiency can be seen by companies.

How can you maximize AI?

Because the AI development will affect many jobs and jobs, it's also important to look at some of the issues that AI could bring with it.

* By identifying an efficient process, the solution for the bias issue in AI is expected to be found through the company.

* It is essential to ensure through the state that results of AI are equally distributed with the affected individuals and those not affected by the advancements.

The issue must be dealt with in a way that is informative for the best results from AI. The students will be able to be trained to perform AI related jobs by way of education programs.

Thus, significant importance must be accorded to STEM subjects. Furthermore, we need to encourage students to develop their creativity and emotional capabilities. Even though, when compared to human beings artificial intelligence can be efficient, humans are successful in completing tasks better than machines in all areas in which relationship building and resiliency are required.

The workplace and the outside of work will soon be altered by artificial intelligence. Instead of focusing the fear of automation technology, it is essential that the new technologies are readily accepted by businesses to ensure that the effective AI systems are utilized to improve and enhance human intelligence.

Making sure that machines don't take Over

If the possibility of a some technology taking over the entire world was announced by Stephen Hawking, it was logical to think about the possibility of it.

The year 2014 was the first time Hawking said that humans may end due to AI. The world has seen numerous benefitsdue to the fact that human intelligence is among the components that are uncovered in each breakthrough. The prediction of the achievements and milestones we can expect to be able to achieve following the implementation of these AI technology isn't an easy thing to do however, we can't undermine the fight against the ailment and poverty.

For a long time, the we have witnessed humanity's possible loss of life at the hands machines. The March 2017 economic report outlines and predicts that machines could take over more than 30% of UK jobs in the next decade. People at risk could include those who work in the transportation, storage manufacturing, retail and manufacturing sectors.

There are a few who have an opinion that is different from the chilling findings of PWC. For example, Accenture estimates that an additional $814billion could be made in the UK economy due to the impact of AI technology.

A myriad of terrible and horrible stories have been written concerning AI and its effects of displacement on jobs has been the subject of numerous studies. Although, this may be seen as a simplistic notion from the viewpoint of many.

The GDP isn't getting boosted by conventional boosters of financial growth but the bright rays are brought by AI.

In this respect, experts believe that a new type of virtual work is AI that is able to overtake the traditional market and change the underlying causes of growth.

Artificial intelligence and automation are encouraged by AI. They are two distinct concepts.

The first involves the use of data to provide services, as well as the intelligent implementation of task. In contrast, the latter recognizes the tasks and allows us to complete the tasks in a way that is efficient.

The increased availability of cloud computing and the rise in the cost of computing power has spurred AI in the field of business. Governments and companies around the world have embraced cloud-based platforms to analyse the massive collection of data in near-real time.

Every bit of information can be stored and merged by these algorithms to provide an exciting and new possibility.

It is difficult to convince your top managers to move forward by involving a math professor is needed to aid you in explaining ML or AI algorithms.

Through the development of applications based on the basic technology This has been done by Ayasdi however the main objective was to focus on business issues, such as discovering the most effective methods in healthcare based on hospital records so that the highest quality of care can be provided at a reasonable cost.

AI companies are already offering assistance to businesses to become more efficient and intelligent. The AI technologies have the potential to create enormous financial benefits, but it is unclear how do the infrastructure and human resources can be built and trained in order to achieve the financial benefits.

In the near future, traditional occupations will be replaced by technologically driven jobs, in which you'll likely be leading machines or doing work by machines. This is why we need to train the people to prepare for the new challenges that are coming.

Things to Consider about the Future of AI

1. AI is significantly expanding beyond our wildest dreams

2. Artificial Intelligence is used being used every day by many people

A few of the most common examples of AI include Cortana or Siri. AI is gaining traction all over the world. Video games, automobiles and lawnmowers, as well as vacuum cleaners are all witnessing the usage of AI. Additionally, a few other examples could include the international markets for financial services, software for E-commerce and medical research areas.

3. Robots can take over a portion of Your Job

You may be showing an excellent performance in your work. However, the majority of routines in the office have been automated or are scheduled to be automated in due course. According to Professor Moshe Vardi at Rice University Rice University, most jobs are going to be completed by robots in the next three years. This may not sound like a good idea however many scientists believe it will be a fresh start, with work being performed with joy and not in the absence of need.

4. A lot of Intelligent People believe Building AI to a Human Level is a risky task to undertake.

There are many disturbing things that occur when machines behave as humans. There is a small chance that the growth of AI will be stopped in the near future.

5. If AI becomes smarter than us It is a bleak Chance of Learning it

Super intelligent devices work ahead of our thinking. It could take many years to become acquainted into the simplest things we are experienced with however, there is no tangible results.

6. Super Intelligent AI might work through Three Techniques

7. The reason why we haven't Meet Aliens Could be None Other than AI

Intelligent Future

It is not too long before the day arrives where the jobs of the past are going to be replaced by intelligent machines, paired with robotics that look like humans, and the high performance of the latest computing technology.

However, in certain situations, AI programs deliver better output than human race, for example, patterns recognition tasks or engaging in video game. These are just the repetitive tasks that machines can perform better than humans, and humans are unable to keep up with AI in these scenarios.

It took decades and years by the human brain to transform from basic things to technological advancements, while also performing productive tasks. The human brain and the intelligence of machines are not able to operate in the same direction.

We can understand the processing of information from basic animal brains. For example, even though they lack an forebrain level of intelligence is displayed by small animals such as octopus and honeybees.

When compared to AI and AI, the efficiency that these creatures have is higher because they are able to perform a variety of tasks spending a minimum amount of time. In the end, millions of samples are required by deep-learning networks before they can learn.

To summarize AI is growing in recognition across the majority of the fields. For instance, the growth rate of AI is reported to be higher in the last year. In recent times, it has become more common to recognize that AI is recognized by a lot of people due to its benefits in the professional and personal lives of human beings. The main advantage for AI is the fact that it can provide human-like outputs as well as reducing costs, cutting time, and providing extraordinary experiences.

Chapter 11: Chatbots And Autoresponders

AI driven by digital solutions that improve services to customers is the basis of a revolutionary technology, and every industry is working on to revolutionize customer service across all areas, such as product knowledge, branding awareness, customer acquisition loyalty programs, as well as after-sales services.

What are Chatbots?

Chatbots are a type of Artificial Intelligence (AI), that is an automated robot system that can create virtual conversations using messages, text chats or both, just as human conversations, but for example, to provide automated responses or perform particular tasks that users need to complete, based on an established set of guidelines or rules.

Second, chatbots is a chatbot that can be utilized on websites, messaging platforms or apps that allow it to fulfill a variety of functions.

Chatbot is also known by the name of Talkbot as well as Bot in short, can be described by its definition as Artificial Intelligence computer program or an application that is simplified in software, which is designed to emulate or simulate human conversation or interactions with people, generally via the internet in a controlled method.

The increasing demand and popularity of these messaging apps has outsmarted even the most well-known social media platforms that date up to the year of 2015. This means that as the popularity for messaging Apps grows increasing numbers of consumers are likely to want to use chats for interactions with brands and companies. In the year 2017 the number of conversations that automated users can participate in has increased dramatically, changing the way brands interact with customers. One illustration could be Facebook Messenger, which in the year 2016 beat the record for active users by more than 990 million.

Sales and customer service for businesses will be greatly improved by the utilization of the widely used chatbots that communicate, resulting in increased customer engagement and higher profits and earnings

What is a Chatbot and how does it work?

Perhaps you are curious as many others are about how chatbots function. It is vital to know this because the current trends indicate that in the coming years, this element of automation will play a significant role , especially in the field of business. If you are a businessperson it is possible to experiment with this when the planning of your business to ensure it is in line with your exact business requirements.

Two kinds of chatbot software programs in current times comprise Artificial Intelligence Chatbots that process natural languages. This allows the chatbot to understand and interpret human speech or text messages as well as discern the intent. One example is when

If you send a message asking for something when you send a message, when you send a message, the system can be "smart" capable enough detect and interpret the meaning. It also responds with questions like "do you prefer coffee or tea?"

In contrast the rule-based chatbot is where specific commands are employed to get the response. This chatbot type is an opportunity to improve customer service and increased profits such as the use of a text messaging app that is used by certain retail companies to communicate offers or coupon coupons that are advertised by an advertisement on the coffee shop that reads "Text COFFEE 21534 for the 5% discount."

Another scenario is that if your phone text "COFFEE" via its messaging service, then the app detects the word as a request, and adheres to the developer's rule and immediately gives customers the voucher code. If , however, you type in the words "COFFEE" is not correctly spelled, it may not be picked.

Chatbots could be considered to be an Application that does not have a User Interface. To get a better understanding, let's look at the functions of an online store in comparison to Chatbots. Chatbot. The latter is a computer application that sells items. The layers or elements which make it function are The following elements:

An Application layer A set of instructions that function as a basis for the APP.

App programming interfaces (APIs) connect the App to services , so as to , for example, payment or shipping quotations.

User interface: Allows users to tell the application of what his main interest is or what he would like to do.

Database: Stores customer information as well as product information, the bulk of the information and transactions.

The user interface on the website for e-commerce is an essential element of the functionality of the app. It includes, for instance, an advanced search tool that can be used to locate products for shoppers while and at the same time, showing stunning product images. It includes user buttons that connect to shopping carts, where you can add items. There are templates to input addresses and payment information.

The chatbot (chatbot) is powered by natural processing of languages and Artificial Intelligence includes three of the aspects of e-commerce that include the application layer, the database and accessibility to Application Programming Interfaces (APIs). However, it does not possess its own user interface however it relies on the messaging platform. It could for instance utilize Slack, Facebook Messenger, WhatsApp or another similar App to facilitate customer or shopper interaction. It is the case that developers will need to create links to other platforms mentioned earlier in order to possibly achieve the same reach as an online store.

What is an autoresponder?

An Autoresponder is an automated service that allows users to send out emails to various groups, or even a particular group of individuals.

It's also a powerful yet easy marketing tool that is utilized to send a sequence of scheduled messages based upon particular criteria. This can be used to the purpose of following up with customers or shoppers within a retail store or company.

For the purpose of promoting specific items and also for requirements, An email could be sent to all clients or customers, for instance to follow-up. This could take the shape of a voucher or specific coupon, or simply to inform customers of similar items or accessories that are in perfect harmony with the item they bought.

What are the functions of Autoresponders?

Autoresponders are an effective marketing tool that's efficient and vital for online entrepreneurs and marketers. They let you contact numerous potential customers. When you consider the number of emails you'll need to send to achieve an individual sale, using an autoresponder is an excellent benefit for your company. Autoresponders can automate up to 50% of a marketing strategy, without which businesses could miss out on the majority of sales per year.

Autoresponders and email marketing tools are extremely important business tools according to the following:

Automatic email notification

Automatic email response

It is used to follow up with customers.

Contacting clients with information about products for example. price lists and special offers or newsletters

To increase loyalty

to keep visitors connected to your website.

To generate sales.

Development of email courses

After email messages are written then they are sent on autopilot, and then subscribers sign up to receive them.

Certain autoresponders that are commercial along with sending out standard messages, are able to send unlimited follow-up messages to members immediately, with no time intervals.

Certain hosting providers can offer an autoresponder at no cost but, purchasing an independent service provides an additional scope to perform additional tasks, such as personalizing emails. One could include the name, telephone numbers, addresses, or business name, and even the attachment of an official business card.

161

Autoresponders allow the creation of short emails that contain pertinent details, lasting for 10 to 15 days that can be provided to your visitors for no cost. This will attract targeted customers based on the subject chosen. Article briefs are sent automatically daily to visitors. This also builds a name for the business.

Some examples of Autoresponder tools that you can investigate include:

* Office Auto Pilot

* Campaign Monitor

* Aweber (used to create and send emails for marketing purposes)

* Infusion Soft

* Auto Response Plus (ARP Reach)

* Mail Chimp

* Constant Contact

* Get a Response

* iContact

* Shopping Cart

Persuasion by AI automation, data science

Persuasion is defined as persuading. It is a form of influence or incitement. One can persuade through attempting to influence their actions or attitudes, motivations or beliefs, as well as intentions. In the business world, however the method of persuasion is intended to influence or altering the behavior or behavior of a group or individual toward an concept, object, event or even towards another person through spoken, visual or written language to communicate information, reasoning and feelings or a mixture of all three.

Persuasion is a technique which is frequently employed to gain personal advantage like campaigning for a political position, the field of sales promotion or advocacy. Persuasion could also be defined as the process of using one's status or position to influence the behavior or attitudes of people.

Heuristic persuasion occurs when opinions or beliefs are influenced through emotion or habit.

The process of persuasion by which beliefs or attitudes are influenced by logic and reason.

Automation is a technology-based procedure or procedure that is carried out with minimal human intervention. It is also known as automatic control, it involves a range of control systems for the operation of equipment. Equipment such as machinery factories, ovens, processes in factories for heating treatment boilers for telephone networks as well as stabilization and steering of vessels and aircrafts, vehicles that have minimal or no human intervention , and many other applications.

Automation encompasses applications that range from huge industrial control systems, to the household boilers that are controlled by thermostats and from advanced multi-variable algorithms, to simple controls that turn off.

Automation is often used as a combination of pneumatic, mechanical, hydraulic electronic, and computers. Complex systems, such as ships, airplanes and modern factories typically employ these methods. Automation benefits include lower cost of labor, savings on material expenses and electricity expenses as well as improvements in accuracy as well as quality and precision.

Scientists assist in the retrieval of useful data from the sea of data that needs to be analyzed and then use the findings to improve the efficiency of companies. Data scientists' role is to study data and ask questions that are based on data employing mathematics and statistical methods to uncover crucial results.

In this digital age in which there is an astonishing growth in automation and automation, there is a pressing necessity to start dialogue and conversation, and especially in the current times when we're confronted by the fascinating area of Artificial intelligence. There are intelligent voice chatbots, autoresponders, Netflix, self-driving cars and the robotics field to mention just the few. To be able to accept Artificial Intelligence (AI) for instance, we need to be educated on not just how the various components work, but also how it impacts society. The majority of arguments about automation are primarily focused around the employment market and its positive or negative effects, especially when AI systems are becoming more widespread.

Two ways that present trends in automation will impact and influence our future is that it will without doubt create an efficient and better future, as evidenced by the industrial revolution. Jobs will be lost but because of the rapid growth the chances of openings will increase to fill more modern, advanced and productive jobs.

The alternative is to acknowledge that the current time is a new one as robots become ever more efficient and intelligent. It is likely that the number of employment opportunities and companies they devour or destroy will be greater than by a significant amount the number of jobs they create.

If either of the two scenarios above turns out to be an actuality What we do know with certainty is that the rise of automation or AI for that matter is dramatically impacting the economy in these incredibly challenging times. We are now realizing that a large percentage or part of our jobs and lives are being increasingly automated.

This issue is not new since in the 19th century, the workers of textiles in England have been documented as having destroyed weaving looms they were later accused of being against technological advances.

The discussion is full of the possibility of intelligent machines in the future times turning against humans or their creators in the end. The possibility of disruption to market for jobs by machines that are intelligent has been studied in academic research and various publications that have been published recently. In addition, thorough studies have been conducted concerning the risk posed by automation to human life in general and specifically with regard to AI.

The AI technology was not predicted more accurately than in the classic science-fiction Space Odyssey film of 1968 that was written and directed by Stanley Kubrick. The film, which follows an expedition to Jupiter finds underneath the Lunar surface, a mysterious featureless artifact , or alien monolith. Mankind embarks for an adventure to uncover the origin of his existence using HAL 9000, a highly intelligent and sentient (perceptive and smart) supercomputer.

Automation is not a new concept in the sense that robots have been used for many years in factories and assembly lines. The current wave of automation derives its advantages from low-cost computing power and the emergence of new applications in software like processing images and language. The latest generation of automation devices and AI technology could negatively impact jobs in the white collar that wasn't previously the case. Beyond the possible impact to the job market, there could be broad implications in other areas like education privacy, cybersecurity health as well as environmental management and energy.

The cost of bandwidth, for instance, can affect learning methods, which is evident via online platforms that are significant in influencing how and what that is learned. One example of this is MOOCs (Massive Open Online Courses). As automation takes on the majority of everyday tasks the kind of education that stimulates conceptual and creative capacities is becoming more popular and well-known, leading to a change in the educational system away from traditional mathematics and reading to more focused on personal and intellectual abilities, which work closely with intelligent machines similar to AI.

All-encompassing commercial and personal data storage and data collection by social media platforms, on the other hand creates privacy issues. Another risk area is cyber-security. The advanced economies are experiencing the effects of automation more so than the emerging countries, which are closely, particularly by investors.

Customer service: Reduce the workload and handle retention as well as segmentation, scoring and score

Two advantages of using AI for customer support are reduced workload and increased customer retention.

There is a growing demand worldwide for superior customer service as a result of increasing exposure to newer technology and solutions. The expectations of customers are increasing. the demand for seamless services is increasing every day, particularly with increasing automation of services that are aided by technology with messaging platforms, social platforms and apps.

Retention of customers is crucial because it increases sales, and the aim of business to increase loyalty and thus keep customers. It's no surprise that a good service to customers is the key to satisfaction of customers, which turns them into brand ambassadors. They offer promotions for free to family and friends, thus increasing sales and thereby increasing revenues.

It has been proven beyond doubt that bad service can turn away customers, but yet, they come back when they are treated with respect regularly. Customers are more apt to use sophisticated and effective systems, and they are unable to choose an outdated, inefficient technology when they have alternatives modern channels that are more efficient and efficient, similar to automated systems like AI are able to provide.

Artificial Intelligence (AI) therefore integrates intelligent services and automation to attain the highest level of accuracy, efficiency, and scalability that manpower alone is not able to attain.

AI can be trained, tuned and customized to match business processes of the present time. Customer service that is AI-enabled through contact centers increase or improve human agents' capabilities. AI chatbots to provide customers with a pleasant service experience. AI chatbots can answer simple customer questions and provide solutions, while human agents are able to manage more complicated issues.

Predicting the behaviour of consumers and referral programs

The emergence in Artificial Intelligence (AI) could assist in the understanding of the needs and desires of consumers, prior to them even establishing themselves. Deep learning, which is a subset of AI can change marketing by assisting in the prediction of consumer behaviour by businesses. This method of machine learning employs layers of neural networks that are similar

171

to the brains of biological species that can not just acquire or develop skills , but also to solve difficult problems at a greater speed than humans. This helps computers or robots to complete "human" tasks such as hearing speech, translating language and recognizing objects. With inputs, deep-learning can to train AI in the process of predicting outputs. But deep learning needs a lot of computation and data, however, less data processing by humans is needed when compared to machine learning methods. With access to the most important elements deep learning systems can be trained to predict human behavior quite precise.

Referral programs are now common methods to acquire customers. But there's nothing to establish that these kinds of customers have more value over regular clients. There is a debate about whether those who are referred to customers are more loyal and successful and, if so how much? Researchers have tried to answer this.

A major German bank was surveyed by 10,000 customers over a time of approximately three years. The results showed that the bank referred customers:

1). Display a greater contribution margin rate, which eroded by slow

2). Indicate a higher level of retention than regular customers who remain loyal for a long time

3). In the short and long term they are superior to the rival customer with the same date for acquisition as well as demographics.

A customer who was referred had a an average of 16% more than the non-referred customer.

Chapter 12: The Right Artificial Internet

Affectations For Your Business

It is crucial to know the ways that Artificial Intelligence can have beneficial for your company. There are a variety of AI applications which can be employed within businesses to make the running of your business much more efficient. There are a variety of Ai applications that, when properly utilized can play a significant part in the transformation of your company.

Automated systems incorporated with AI will greatly assist you make better choices in the management of your company's resources. There are a variety of Artificial Intelligence applications that can be utilized in the business world to benefit the business. A few of the Artificial intelligence applications are utilized across various sectors, such as healthcare, customer service financial security, drones, intelligent cars, creative arts among others.

Today today, the majority of customer concerns or questions are attended to by humans; it is accomplished via phone calls or emails, as well as

174

chats online. Yet, AI applications have been implemented in corporate systems, which can facilitate automatization of these communication. Computers can be incorporated using Artificial Intelligence applications to enable them to provide accurate responses to clients. Furthermore, the combination of AI and ML allows the AI machines and computers to function more effectively. Deciding whether or not to implement AI for your business will depend on the goals you intend to accomplish over the long term.

How can artificial intelligence (AI) applications be utilized in accounting for the field of business?

Accounting for business is an organized recording of, analysis as well as the interpretation as well as presentation of financial data. For small-scale businesses, accounting is usually managed by a single person, or in a variety of teams within larger companies.

Accounting is of major importance to any business because it assists the owner of the business to keep track of the company's activities. Accounting processes aid in analyzing financials, which allows the business owner to make better choices, particularly in managing finances. Implementing Artificial Intelligence applications in businesses helps with accounting for business owners, allowing owners to meet their obligations to comply.

Artificial Intelligence software is set to revolutionize the financial and accounting industries. they will eliminate repetitive tasks, which results in savings in time. Employees can thus be engaged in other jobs that have a greater effective impact on the company.

HANA can be described as an Artificial Intelligence application that helps companies convert their databases into tools for intelligence which are more effective and useful. It works by taking in and replicating data on sales transactions taken from relational software and databases.

To gain deeper insight and streamlining processes in finance, businesses must look into current AI applications. These programs help accountants to stay on top of business transactions in lengthy and tiring procedures. This is possible through using Machine Learning Machine Learning app instead of PDFs and spreadsheets. Receipts images are extracted, and classified automatically through the Machine Learning application. The collected data is classified based on the category of spending. These reports provide valuable insights for companies that aid in more efficient financial planning.

The ML application is a subset of AI and provides deeper understanding of how the processing of data takes place in a continuous manner. This gives businesses a complete overview of spending patterns that can be used to make better financial decisions for businesses. The most basic AI applications are definitely important gamers in the accounting and business area.

Do you know if Artificial Intelligence Applications used in the funding of established and startup firms?

Startups have been controlled by various software giants , including SAP, Oracles, IBM and Salesforce. They offer various software which are used for the areas of customer relationship management, resource plan, and the management of human resources. A number of startups are providing the next generation of new services through Artificial Intelligence apps,

Businesses that are driven through Artificial Intelligence, many of which are AI startups, are employing these technologies in their recruitment process. For instance, Salesforce has made investments in DigitalGenius, which is a solution for managing for clients and Unabel which provides business translation services.

Artificial Intelligence startups are now creating solutions for various industries. This is evident across many business areas i.e. agricultural, legal, automotive health, finance and healthcare. It's simple

These Artificial Intelligence startups are becoming extremely popular due to their ability to supply established companies with sought-after points solutions. This is because they are able to access domain knowledge as well as large volumes of data.

The large majority AI startups are currently developing the same type of Machine Learning models that have the capability to predict or categorize end-products using specific input data. For the vast majority of AI startups, the effectiveness of running is heavily dependent on volume of data; the greater the volume of data greater their efficiency.

Artificial Intelligence in Creating the most effective team to ensure the way to success in business

The rapid growth in the use of Artificial intelligence in the business world is a result of the necessity to answer a variety of questions during the process of recruiting i.e. precisely what you're seeking in the candidates you seek to recruit, organization structures that work best for your company the management module that will lead to the success of employing Artificial Intelligence, etc.

Many companies have realized the significance of adopting Artificial Intelligence teams that play significant roles in enhancing efficiency and delivering services to customers.

Utilizing AI teams with a broad range of expertise will allow businesses to maximize AI benefits.

When it comes to choosing the members of your Artificial Intelligence team, it's essential to think about reorganizing your hiring processes.

It is essential to have a skill set who can comprehend the computer-based learning process. A central team is the driving force behind Artificial Intelligence applications. Remember that the primary goal in Artificial Intelligence is to basically reduce the amount of work to be done in the shortest time. A well organized team is essential to ensure greater efficiency of the company that will reflect positively in the future.

The impact of is having the best team within any organization can never be undervalued. This is the reason why employers must to make use of the beneficial AI applications to allow them to hire those with the proper skills and ensure that work is accomplished in shorter times that can significantly assist in achieving goals and increasing productivity.

This Artificial Intelligence team is tasked with the task of executing the entire AI project with success. This implies that the team needs to have the engineering skills to accomplish this job efficiently. Based on the type of project to be completed Data engineers and data scientists, machine-learning engineers and product managers might be part of members of the AI teams members.

Baidu along with Google are both excellent technology teams. They have demonstrated the capacity to collaborate across functions to create powerful AI value through advertisements and product recommendations speech recognition, etc.

What are Artificial Intelligence applications used in the preparation, collection for recycling and reuse of information?

Data is created in massive quantities in the world of business. They can be categorical, free text or numerical data. Data collection is defined as the process of collecting data and analyzing it using many different sources. If we want to use the data we collect for business problem solving it is essential to have a sensible data collection and storage of data. This will help in the development and implementation of Machine Learning and Artificial Intelligence solutions.

The process of gathering data and preparation is vital in the business world because it allows you to record recent events. Data analysis is then initiated to look for patterns that are recurring. Predictive models are built using these specific patterns that are used to forecast potential changes to come in the future.

It is essential to follow the best practices followed when collecting data to aid in the creation of highly efficient models. This is dependent on the information that is taken. The data collected should be completely free of errors and include relevant data to help in the accomplishment of the task. The data collected is utilized to formulate strategies to solve similar problems or other tasks in the future.

Conclusion

The concept of mimicking the human brain with the aid of symbols and operations is the foundation of Artificial Intelligence. Apart from being a fact Artificial Intelligence can lead various enterprises to a higher level but there's the question of whether Artificial Intelligence can replace human intelligence and act like it. To reap the advantages of AI and AI, it is important to not be waiting to see the outcome of these issues.

There was a period when technology was utilized to automatize procedures and compute business theories. Now, the more crucial thing is the vision and how that vision can be used to create value and achieving the desired goals in the midst of a competitive environments. This can all be accomplished by using Artificial Intelligence. With the aid of AI chances of getting to the top spot in the current business environment will be limitless. Soon every sector will be able seeking assistance with AI as well as ML to improve their business models and efficient. You can see clearly that AI is poised to revolutionize how businesses are conducted.

It is essential to have a plan when you are looking to purchase Artificial Intelligence for your business. However, you must remove the traditional thinking and equip yourself with the latest approach for your business and also for the AI application. This book is designed to serve exactly the same function. It provides you with an overview of the most recent techniques and methods that aid you in analyzing the processes of your company efficiently. Additionally, you will be taught about the various strategies that are the most appropriate for your company's needs. It is important to mention that various industries such as manufacturing and banking, sports and retail have already begun to benefit by AI and machine learning.